RISK MANAGEMENT
FOR DESIGN AND CONSTRUCTION

RISK MANAGEMENT
FOR DESIGN AND CONSTRUCTION

OVIDIU CRETU

ROBERT STEWART

TERRY BERENDS

WILEY

John Wiley & Sons, Inc.

Library of Congress Cataloging-in-Publication Data:

Cretu, Ovidiu.
  Risk management for design and construction / Ovidiu Cretu, Robert Stewart, Terry Berends.
    p. cm.
  Includes bibliographical references and index.
  ISBN 978-0-470-63538-4 (hardback); 978-1-118-02861-2 (ebk.); 978-1-118-02862-9 (ebk.); 978-1-118-02863-6 (ebk.)
1. Building—Safety measures. 2. Construction industry—Accidents—Prevention. 3. Construction industry—Risk management. I. Stewart, Robert, 1968- II. Berends, Terry. III. Title.
  TH443.C74 2011
  690′.22—dc22
                                                                              2010047231

10 9 8 7 6 5 4 3 2 1

First and foremost, I'd like to thank my wife, Tamara, for enthusiastically supporting me in all my activities, and my children, Petru and Vlad, for being there and putting up with me.

—Ovidiu

I wish to dedicate this book to my loving wife Judi and to our children, Ty and Charli, for their support and love.

—Terry

Thanks to my girls, Vanessa and Daphne, for heightening my sensitivity to risk.

—Rob

For all of our colleagues in project management who share our passion to build better values.

—All

CONTENTS

PREFACE

R isk management is perhaps the hottest topic of discussion for professionals within the design and construction industry who care about the fate of their endeavors, whatever they may be. Increasingly, professionals are engaged in risk management even before the project is assigned to them. It is difficult to imagine project management without formal or informal risk management. The next paragraph presents one more reason why risk management is needed for projects that we want to succeed.

> The need to make strategic long-term investment decisions under short-term budget constraints continues to force states to consider risk as a criterion for judging a course of action. Because perceptions of risk vary, decisions incorporating risk management concepts depend to a large extent on the decision maker's tolerance for risk. It is this fact that makes implementing concepts incorporating risk more difficult because it requires decision makers to establish boundaries of acceptability based on political, economic, and engineering constraints, most of which are unknown. Having clear, well defined, risk analysis methods and procedures is a first step toward helping agency leaders begin to incorporate risk into their decision making process.

> *Michael Smith*
> Federal Highway Administration
> michael.smith@dot.gov

Whether we acknowledge it or not, risk management is part of our daily life. For the most part, we assess it on an unconscious level, usually out of habit. For example, consider the ubiquitous nature of the flu. We all know that when the flu strikes, suffering will follow. The fact that a person may or may not be infected by the flu is anyone's guess; however, based on historical data, we may say that there is about a 75 percent chance of being exposed to the flu. Once a person is exposed to the flu, if that person is vaccinated against the flu, he or she still has a 10 percent chance of getting sick, while a person who hasn't been vaccinated has a much higher chance (85 percent) of getting sick (the percentages are simple assumptions).

We know that people manage the risk of getting the flu differently. Spending a few dollars for a flu shot reduces the chance of being infected to 7.5 percent. Within the context of risk management, when a person gets immunized we call this action risk mitigation, because it reduces, but does not eliminate, the chance of getting infected. Another person may elect not to get a flu shot, and he/she is accepting the chance of 67.5 percent of getting sick. This inaction is referred to as risk acceptance, when a person elects to take action (medication) after being infected. A third person may elect to avoid leaving his or her house and remain completely isolated from other people. His or her chance of getting sick is close to zero. In this scenario, the decision to take extreme action to ensure that she/he is not going to be infected is called risk avoidance.

The flu example may be expanded further if we consider and compare the effect of the flu on vaccinated persons versus nonvaccinated ones. The concept of risk impact or consequence explains the second part of the flu story. The flu's impact on a person may be measured by how much that person is going to suffer and for how long. As a general rule, a vaccinated person will suffer less than a nonvaccinated person and will also recover in a shorter time period.

The "suffering" may be measured by the time lost from regular activities, the cost of medicine, infecting others, and so on, and can be expressed through a range. Now things are becoming more complicated so we will end the flu story here since we do not want to expose the book's secret just yet.

The decision of whether or not to get vaccinated is relatively simple; people usually consider all of the facts presented, but do not perform a formal risk analysis. Common sense dictates to each person how to manage the risk of getting the flu and, more often than not, their decision is right.

There is an abundance of sophisticated software available on the market today that allows risk to be quantified. Aside from being expensive, they require specialized training and experience in setting up the models. Further, as most of these are designed to address a wide range of risk modeling scenarios, they include additional features and layers of complexity that are not needed for construction applications. A good model should be as simple as it needs to be and no simpler. Finding this "sweet spot" is important in both building the model and understanding the output.

To paraphrase Alan Davis, author of *Software Requirements—Analysis & Specification,* a model simply provides us with a richer, higher level, and more semantically precise set of elements than the underlying wholesome guesstimate. Using such a model reduces ambiguity, makes it easier to check for incompleteness, and may at times improve understandability.

This book provides a practical framework for managing risk on construction projects, from their inception at the earliest stage of design through the end of construction. A central element in the presentation of this material is risk modeling software that will allow users a simple means of quantifying project risks in terms of their impacts to both time and cost. The software is driven by MS Excel, and does not require any prior knowledge of programming or risk modeling. The information provided in this book is sufficient for any Excel user to get started, run it, and understand and explain the model results.

The next paragraph presents one user's experience with the software described by the book:

> I have used the software multiple times for small to medium size projects. It has been very reliable and provides the ability for the client to continue risk management and updating of the results as the project progresses.

> *Ken L. Smith, PE,* CVS National Director Value Engineering
> Vice President HDR ONE COMPANY | Many Solutions
> ken.l.smith@hdrinc.com

While this book addresses risk in the context of transportation projects, the process and tools presented in it are applicable to any kind of project. The process presented in this book is one of the most versatile methods of estimating and risk assessment and analysis. It can be applied to projects of any size; any stage of development; and any level of complexity. It is just a matter of understanding the project and matching the level of effort to the objective.

Next we present the testimony of one professional who creatively and efficiently applied the process and the tool (Risk-Based Estimate Self-Modeling, RBES) presented in our book.

> The Cathedral Hill Hospital Project, San Francisco, for Sutter Health and California Pacific Medical Center is a $1.7 billion program which includes about $1 billion in construction cost depending upon the final selection of scope. Construction is expected to start in 2011.
>
> Sutter chose to use Integrated Lean Project Delivery™ which unites architects, engineers, and contractors in a collaborative partnership of shared risk and reward. The method and relationships within this arrangement are creating new parameters for risk and its mitigation. *Insurance underwriters are studying the potential adjustments to reflect the reduction in risk*.
>
> To better tell the risk story from an insider's view, a team led by John Koga at HerreroBoldt assembled a Risk Assessment Report listing about 700 standard perils along with their potential cost and schedule impacts under this new form of project delivery. They divided the perils into nine Groups and seventy-four Categories.
>
> Borrowing a current version of the RBES Workbook developed by Dr. Ovidiu Cretu and Terry Berends as a template, they ran Monte Carlo simulations to look at the perils within each category. Selecting perils across the Groups, they built combinations of perils to further understand their exposure. The RBES Workbook gave Koga the ability to adjust the settings associated with ranges of probability and risk outcomes.
>
> RBES also provided the ability to choose adjustments for Escalation and Market Conditions. It provided excellent graphic output that would not be available from a standard spreadsheet without additional effort. Furthermore, RBES allows the risk management effort to occur throughout the life of the project, retiring risks as soon as possible and recalculating exposure. The client especially appreciates that the perils have been made more visible and preventive action or mitigating strategies can be developed.
>
> *John Koga CVS-Life, CDT, LEED AP*
> Member AIA, CSI, USGBC, SAVE International Manager, Value and Lean Process
> HerreroBoldt

The primary audience for this book is those involved in the development and delivery of projects and programs. This includes project and program managers, designers, engineers, architects, cost estimators, schedulers, and risk managers, as well as those seated within upper management who have an interest in developing fluency with risk management. Consultants will find this book particularly well-suited to their needs. Students will also find this book useful as it is written in plain English and does not demand any prerequisite skills or experience.

This book comprises seven chapters united by the concept of risk management. Chapter 1 presents a short introduction on risk management and Chapter 2 presents a concise overview of cost and schedule estimating. These two chapters lay the ground for the next five chapters and they should not be seen as in-depth analyses. For example, Chapter 2 presents the basic concepts and definitions related to cost and schedule estimates but we do not expect this book to be viewed as an estimating manual.

The general expectation is that all participants involved in the processes presented in this book are experts in their fields. The estimators are sagacious ones; the schedulers are experienced and understand the difference between critical and noncritical activities; and the subject matter experts (SMEs) are recognized for their expertise and accomplishments.

Recognizing that a professional may have different roles at different stages of the process, this book often assigns different names to the same professional. For example, the *risk lead* (the person who is responsible for managing risk tasks) may be called the *risk elicitor* when he/she conducts risk elicitation activities, or *modeler* when he/she produces the simulation model, or *risk analyst* when he/she analyzes the assessment results and the model outcomes. In other words, we are recognizing that a person may wear different hats at different stages of the process.

Chapter 3 presents the process of cost and schedule risk assessment and analysis which we call the Risk-Based Estimate (RBE). Chapter 3 describes the process of RBE and the main advantages and disadvantages of it. It introduces and defines the concepts of "base uncertainty, risks, and Monte Carlo statistical analysis" as applied to the RBE. This chapter presents lessons learned from our practices and observations of hundreds of projects.

Chapter 3 also presents a new concept of doing risk assessment and risk analysis based on the old saying "Keep It Simple Stupid" with a little twist at its end: "Keep It Simple Smarty" (KISS). The concept "Professional Sophistication" is presented with its pitfalls and dangers. The chapter presents how to apply the RBE on two projects: (1) a simple one and (2) a complex one, in order to demonstrate the process flexibility and applicability.

Joe O'Carroll, an experienced project manager with Parsons Brinckerhoff (PB), familiar with the RBE process, described in this chapter says:

> A fundamental maxim of modern project management is: "If you don't know it, you can't measure it; if you can't measure it, you can't control it." Therefore, controlling cost and schedule overruns has to start with knowing the risks—that is, identifying them early, measuring them or quantifying them by the most appropriate methods, and then managing them. We cannot manage risk that we don't see and won't see if we don't look.
>
> Mapping risk impacts against cost structure, design and construction schedules and against expected project performance exposes risk impacts that have the greatest effect on project budgets, schedules and operational goals. Quantifying risks appropriately can be used to develop targeted contingency funds. This needs to be followed by creation of a project risk and contingency management plan whereby

the drawdown of contingency at key project milestones can be carefully monitored and controlled.

Joe O'Carroll, Risk Manager Parsons Brinckerhoff
OCarroll@PBWorld.com

In other words, risk management is to *know it*, *measure it*, and *control it*. This book focuses on fulfilling all three of these imperatives as they are fundamental to good project management practices.

Chapter 4 focuses on risk elicitation and risk conditionality (dependency and correlation) and discusses the primary means of collecting risk information for the RBE process. It presents the main biases that can alter the value of risks elicited and identifies techniques on how to detect, assuage, or avoid the most detrimental ones.

The next paragraph presents how the program risk manager for the Alaskan Way Viaduct (AWV) project (over US$1 billion) applied techniques to reduce bias during the risk elicitation phase.

> An example of the changes that were made included having at least two of every type of subject matter expert (SME) in the workshop. Previously, in every other workshop I had attended in the past, only one or two independent SMEs were invited to attend risk workshops, and WSDOT employees were invited to be SMEs; this scenario created the potential for (or the perception of) the introduction of unnecessary bias. With multiple independent SMEs in each subject area, we felt that a significant amount of bias towards risk identification and assessment would be avoided.

Steve Warhoe
Past President of the Association for the Advancement of Cost Engineering (AACE)
gongchengshi@comcast.net

This is indeed an effective strategy, if the project can afford to engage multiple SMEs on a critical area. In our opinion, this is money well spent—prevention is always less costly than the cure in the long run.

Chapter 4 continues with a discussion of the elicitation of risk probability of occurrence and impact. The chapter presents and illustrates how risk conditionality affects the results of risk analysis. It begins with risk dependency where the term *risk mesh* is introduced and continues with risk correlation and the combined effect of dependency and correlation. The chapter recognizes the full array of challenges to the risk elicitation process as Susan Adibi wisely states:

> "Elicitation of the project risks and opportunities is arguably one of the most difficult parts of the risk lead role. The risk elicitor needs to have a combination of personable social skills as well as understanding the group psychology and dynamics in order to fully capture the relevant risks."

Susan Adibi, Risk Analyst Parsons Brinckerhoff
adibi@PBWorld.com

Chapter 5 presents an overview of the second pier of risk management (risk response, risk monitoring, and control) by avoiding repetitions of what other books are presenting. A short synopsis of prospect theory and risk tolerance is provided and is tied to the needs of risk response planning, monitoring, and control. The chapter completes the project risk management process and stresses the importance of paying attention to all phases of the risk management process since if one of these phases is weak the process may break down. As Joe O'Carroll stated, we cannot afford to be doing a superficial job when we manage a project.

> Under-managed project risk costs the Engineering and Construction Industry $3–$4 billion annually in profits. It is estimated that, industry wide, $200–$270 billion in additional revenue would be needed to make up for this lost profit' (Datamonitor, Factset; ENR and industry expert estimates of project failure rates). A U.S. Department of Transportation study of 10 rail transport projects found that their capital cost overrun ranged from minus 10% to plus 106%, averaging plus 61%. In a study of 258 projects carried out by Aalborg University, Denmark, it was found that the costs of 9 out of 10 transport infrastructure projects are underestimated, resulting in cost overruns. Furthermore, the study concluded that the cost underestimations and overruns have not decreased over the past seventy years. Only when our industry adopts a total risk management philosophy from concept through closeout of a project do we stand a chance of reversing the tide and consistently managing our mega-projects to a predictable budget and schedule.
>
> *Joe O'Carroll,* Risk Manager Parsons Brinckerhoff
> OCarroll@PBWorld.com

The chapter continues with presenting the integration of Value Engineering (VE) and risk response planning and presents the concept of disaster contingency planning.

Chapter 6 presents the Risk-Based Estimate Self-Modeling (RBES) Spreadsheet which is an MS Excel®–based tool that facilitates the RBE process by employing the Monte Carlo Method for statistical analysis of the data collected through the RBE process. The simplified version of RBES may be downloaded at: www.Cretugroup.com for free, or for a fee the full copy of RBES is available.

The strengths of RBES lie within its versatility and user friendliness that enables users with minimal training to run risk analysis.

> The RBES has the ability to provide results immediately after the scope and risks have been entered. The ability to show real time results is a necessity when a Cost Risk Assessment (CRA) is combined with a Value Engineering (VE) Study. This ability allows the VE Team to target the major risks of any project and speculate on ways to respond to them in a pro-active way. Because the RBES can capture, analyze, and display the results of both "pre-response" and "post-response" scenarios on the

same graph a VE Team can immediately see the impacts that the response strategies and VE recommendations will have.

Ken L. Smith, PE, CVS National Director Value Engineering
Vice President
HDR ONE COMPANY | Many Solutions
ken.l.smith@hdrinc.com

The final chapter of this book is dedicated to explaining how a typical RBE process can be implemented. It presents in a systematic way the workshop logistics; identifies the participants and their roles and responsibilities; and identifies the end products of the entire effort of cost and schedule risk estimating and analysis.

The entire process, once understood and practiced, is appreciated for its down-to-earth development, for its transparency and clarity of data entered, and results provided. Following is a testimony from one experienced estimator and risk manager:

I believe that the RBE process is one of the best risk assessment processes there is; it is an excellent management tool. The process, when followed correctly, creates consistency for project managers, executive management, and other stakeholders in the development of project risk profiles and range forecasts for final cost and completion dates.

Steve Warhoe
Past President of the Association for the Advancement of Cost Engineering (AACE)
gongchengshi@comcast.net

The entire book was designed to provide help to professionals who may want to engage in the practical application of risk assessment, analysis, response planning, monitoring, and control. The book seeks to demystify the notion of the "model" and efforts were made to provide transparency and clarity of the model. The authors believe that when people understand the process they will embrace it and apply it in an appropriate manner.

ACKNOWLEDGMENTS

The authors express their gratitude to the leadership at the Washington State Department of Transportation for supporting and nourishing an innovative and creative culture. We would like to recognize the Secretary of Transportation, Paula Hammond; the Chief Operating Officer, David Dye; and the Chief Engineer, Jerry Lenzy, for their continuous commitment toward enhancing the role of risk management across the organization.

We are grateful to Pasco Bakotich, the State Design Engineer, and his team; Mark Gabel and his group; Laura Peterson, Glenn Wagemann, Amity Trowbridge, Jeff Minnick, Mitch Reister, and Dawn McIntosh for their dedication toward enhancing the practice of risk management and estimating.

We would like to acknowledge Doug MacDonald, Jennifer Brown, John White, Julie Meredith, Ronald Paananen, John Reilly, Mike McBride, Stu Anderson, Keith Molenaar, Dwight Sangrey, Williams Roberts, Travis McGrath, Cliff Mansfield and all others for their contribution to developing a risk awareness culture in the State of Washington and across the nation.

Further, the authors would like to express their gratitude to David Supensky for his inspiring cartoons, Vlad Cretu for his contribution on writing Chapters 3 and 4, Michael Smith for his insights on reviewing Chapter 3, and many others who indirectly shaped the content of this book.

Finally, we want to express our deepest gratitude to our wives Tamara, Judi, and Vanessa, who bravely provided support and encouragement throughout this work.

RISK MANAGEMENT
FOR DESIGN AND CONSTRUCTION

CHAPTER 1

WHY AND WHAT IS RISK MANAGEMENT?

The pessimist sees difficulty in every opportunity. The optimist sees the opportunity in every difficulty.

—Winston Churchill

INTRODUCTION

Risk is something that we deal with in our daily lives. For the most part, we assess it on an unconscious level, usually out of habit. Will I get hit by a car if I cross the street? Will I burn my hand if I take a pan out of the oven without a hot pad? Will I get sunburned today if I don't put on lotion? If you are reading this paragraph now, it is probably safe to assume that you are pretty good at dealing with these mundane risks. Most of us are. This type of risk analysis is intuitive and is built upon years of experience, intuition, and instinct. Generally speaking, the decision making involved is pretty straightforward and seldom do we devote much time to our analysis. It seems that when we face risks personally, they are generally easier to deal with and we arrive at answers rather quickly.

In contrast, when risks are removed from us, they seem to take on additional complexity. This is especially true when considering decisions related to the development and delivery of construction projects. There are countless risks that such projects can encounter at any point of its lifecycle. What if the geotechnical information is wrong and the foundation collapses? What if the price of steel skyrockets two years from now when construction starts? What if it drops? What if the project's environmental document is held up in review and delays the construction bid date? To be certain, the answers to these problems are not always clear. To make matters even murkier, should such events occur, they are likely to spawn a host of other uncertainties. Thinking along these lines is at best challenging and at worst completely overwhelming.

Nassim Nicholas Taleb, one of the most articulate, and arguably the most brilliant, contemporary scholars who writes on the subject of uncertainty, has popularized the term "Black Swan," which he wrote about in a book bearing the same title. A "Black Swan," as defined by Taleb, possesses the following attributes:

> First, it is an outlier, as it lies outside the realm of regular expectations, because nothing in the past can convincingly point to its possibility. Second, it carries an extreme impact. Third, in spite of its outlier status, human nature makes us concoct explanations for its occurrence after the fact, making it explainable and predictable. I stop and summarize the triplet: rarity, extreme impact, and retrospective (though not prospective) predictability. A small number of Black Swans explain almost everything in our world, from the success of ideas and religions, to the dynamics of historical events, to elements of our own personal lives.[1]

To be sure, this book is not focused on managing Black Swan events. However, it would be disingenuous, to say the least, that the contents of this book will enable the reader to avoid, or even manage Black Swans. Therefore, before we go any further, let us acknowledge that there is, and will always be, some degree of uncertainty in all endeavors that extend into the future that lay beyond our perception, whether they are Black Swans or of the more prosaic variety.

We must approach the concept of uncertainty with total honesty if we are to approach it at all. The truth is that it's impossible to predict the future, regardless of the sophistication of the analytical techniques we apply and the expertise of the personnel we bring to bear.

At this point, it is useful to think about the ways in which we should think about the analysis of risk and how this information should be integrated into the decision-making process. Let us call one approach "risk-based" and the other "risk-informed" decision making.

Risk-based decision making is predicated on making decisions solely on the numerical results of a risk assessment. This approach relies on quantitative data to make predictions. Although such an approach is seemingly stochastic in nature, it is in fact deterministic, because it must make the assumption that the analysis that generated the quantitative data is all inclusive, which, as discussed in the previous paragraphs, is not possible.

In contrast, risk-informed decision making is based on synthesizing the quantitative data gained from risk analysis, along with other factors such as the anticipated benefits, functional performance, and political considerations, to name a few. This approach relies on using quantitative data to develop insights that will lead to improved decision making in the face of uncertainty. The quantitative data derived from risk analysis is not the sole basis for making decisions, but is critical information that must be considered within the context of the project and with a full appreciation of its limitations.

In summary, this book is about helping project owners, managers, and design professionals improve project value within the framework of risk-informed decision making. The processes and techniques presented in this book will assist readers to better understand and manage risk by developing insights into the nature and composition of project risks. Specifically, much of this book focuses on the process of comprehensive risk assessment. The approach to risk assessment that will be presented herein is referred to as risk-based estimating (RBE), which provides a method of quantifying project uncertainty, including risk events, and expressing its potential range of impact to a project's cost and schedule.

WHAT IS AN ESTIMATE?

What exactly is an estimate? This is an important question to ask, and the answer may surprise some. The American Heritage Dictionary defines an estimate (i.e., noun form) as follows:

1. The act of evaluating or appraising.
2. A tentative evaluation or rough calculation, as of worth, quantity, or size.
3. A statement of the approximate cost of work to be done, such as a building project or car repairs.
4. A judgment based on one's impressions; an opinion.

Key words to note in these definitions include: tentative, rough, approximate, judgment, opinion. In other words, an estimate is not a precise number, but rather, an approximate judgment of what the actual costs or time will be. In fact, it is not uncommon for professional construction cost estimators to refer to an estimate as "an opinion of cost." This definition may indeed be in keeping with your understanding of what an estimate is; however, it would seem that this is not how most of us actually think about estimates in real life.

There seems to be something magical about the act of printing a number, any number, on a piece of paper that somehow conveys to us that it is a fact, whether it really is or not. For example, when you go to the mechanic to get your car fixed you typically receive an estimate as to the cost of the repairs. You "know" that this is just the mechanic's best guess; however, in practice you tend to take it for granted that it is factual. When you return to pick up your car from the garage, you find that the price has increased significantly, and you tend to feel cheated. "But that's not what the estimate said!" you cry out in shock. "Well, once I removed the carburetor to get to the transmission, I found out that the head gasket was leaking, which damaged the catalytic converter. I ended up having to replace that. I also noticed that the fan belt was shot and that you also needed an oil change. You really ought to take better care of your vehicle!" says the mechanic, his overalls covered in grease. It is easy to see how a $200 repair turned into a $1,000 repair very quickly. We can only accurately estimate what we know. Uncertainty will always throw us for a loop and turn even the most careful estimate into a bad joke.

Executive management, politicians, and the public in general seem to suffer from a similar bias when it comes to dealing with cost estimates on construction projects. "The deal was that the project was estimated to cost $10 million. Now you are telling me its $20 million! What happened!? Are you incompetent?" Anyone who has ever prepared an estimate or received one can probably relate to this scenario.

It must be acknowledged that there is estimating uncertainty and then there is event uncertainty. These two elements are independent, but together comprise the risk profile of a project. We place a lot of weight on estimates, probably more than we should. And yet, it's really the only way we have of planning for future expenditures or timelines that are subject to uncertainty. Therefore, we must have a way to deal with uncertainty in our estimates that allows us to think stochastically rather than deterministically in a manner consistent with the framework of risk-informed decision making.

WHAT IS UNCERTAINTY?

Uncertainty is defined as the quality or state of being uncertain. That is to say, it is a state of not knowing. Within the context of this book, the term *uncertainty* refers to a lack of knowledge about current and future information and circumstances. Uncertainty poses a special set of problems to the management of projects as it can potentially affect outcomes for both the good and the bad.

WHAT IS RISK?

The definition of the term *risk* merits some discussion. It is often assumed that the word *risk* implies a negative outcome. For example, if someone said to me, "That is a very risky assumption," I would take it to mean that she thinks that my assumption is likely to be wrong and, consequently, something bad will happen as a result. The fact of the matter is that risk represents an uncertain outcome. Risks may have either positive or negative outcomes. A negative risk is defined as a *threat* while a positive risk is defined as an *opportunity*. Therefore, something that is properly defined as *risky* does not necessarily mean that it is a bad thing, only that it is an uncertain thing.

This bias toward risks as being bad things often causes us to overlook potential opportunities. Just as threats can result in a catastrophic disaster, opportunities can result in spectacular windfalls. This is the reason for the opening quote for this book. Clearly, Churchill was very shrewd in his ability to perceive both threats and opportunities in employing his wartime strategies. Moreover, he seemed to have possessed the ability to see risks in a balanced way, that is, as both threats and opportunities. Successful risk management requires us to adopt a similar mindset. We must maintain an unbiased outlook and be neither pessimistic nor optimistic in our assessment of risk and be prepared to address both threats and opportunities as they arise.

WHAT IS RISK MANAGEMENT?

A project manager responsible for the delivery of a new office building identifies a permitting concern that could delay the approval of her project. A structural engineer is assessing the quality of the data of a geotechnical report that was performed and fears that the abutments of the bridge he is designing could experience differential settlement. A school district superintendent is concerned that the environmental document could be delayed by public comment. A general contractor fears that the recent volatility in the price of steel could turn a profitable project into a money loser.

All of these scenarios are everyday occurrences within the design and construction industry; however, the manner in which these risks are addressed will have a large impact on project outcomes. The practice of risk management can certainly play an important role in ensuring that the outcomes will be positive ones. However, a lack of risk management will likely result in increases to a project's cost and schedule.

Risk Management is one of the nine Project Management Knowledge Areas identified in *A Guide to the Project Management Body of Knowledge (PMBOK® Guide), Fourth Edition.* The *PMBOK® Guide* is an excellent project management reference published by the Project Management Institute (PMI®).

To quote from the *PMBOK® Guide*, "Risk Management includes the processes concerned with conducting risk management planning, identification, analysis, responses, and monitoring and control on a project. The objectives of project risk management are to increase the probability and impact of positive events and decrease the probability and impact of negative events in the project."[2]

Another definition of risk management provided by the International Organization for Standardization (ISO)[3] identifies the following principles of risk management:

Risk management should:
- create value
- be an integral part of organizational processes
- be part of decision making
- explicitly address uncertainty
- be systematic and structured
- be based on the best available information
- be tailored
- take into account human factors
- be transparent and inclusive
- be dynamic, iterative, and responsive to change
- be capable of continual improvement and enhancement

WHY RISK MANAGEMENT?

Research has shown that historically the majority of construction projects experience cost and/or schedule overruns. A cost overrun is defined as the difference between the low bid and the actual incurred costs at the time of construction completion.

A study focused on analyzing the costs of public works projects in Europe and North America found that the incidence and severity of cost overruns was significantly higher than indicated by the previous source.[4] This same study found that cost overruns were found in 86 percent of the 258 projects that were sampled. Further, actual costs were, on average, 28 percent higher than estimated costs. The authors of the study concluded that the following factors were the primary culprits in cost overruns:

- Lack of proper risk analysis in developing estimates
- Poorly defined scope at the time initial project budgets were developed
- Larger public projects are prone to intentional underestimation due to political pressure (In other words, there was a deliberate misrepresentation of project costs and/or schedule in order to further political agendas.)

Public projects, especially "mega" projects, seem to be especially vulnerable to unfavorable project outcomes resulting from poor risk management. Governments worldwide are making massive capital investments in large infrastructure projects as a means of providing badly needed services while providing a "boost" to flagging economies. In a recent article, *World Finance* reported:

> . . . governments may be throwing these vast sums down the drain if the harsh lessons from other mega infrastructure projects are not learnt. The Channel Tunnel cost double its original budget and only returned a profit 20 years after the project started. Denver's international airport saw its eventual cost triple from what had originally been planned, and Sydney's Opera House—as amazing as it might look—still holds the world record for worst project cost overrun at 1,400 percent over budget. Its construction started in 1959 before either drawings or funds were fully available and when it opened in 1973, 10 years later than the original planned completion date and scaled down considerably, the building had cost A$102m rather than the meager A$7m budgeted.[5]

Clearly, as construction technology becomes more sophisticated and the problems facing society become more complex, the need for prudent risk management will only grow.

THE LIMITATIONS OF RISK MANAGEMENT

To some, the term *risk management* may seem like an oxymoron. How can you manage the future? Further, how can you quantitatively model it? Similarly, why do risk management plans sometimes fail? These are valid questions that merit further discussion.

Taleb has argued that the use of probability distributions, which were first developed by the German mathematician Carl Freidrich Gauss (1777–1855), are inappropriate and were never designed for the purposes of modeling complex future events. The authors agree that there are indeed limitations in applying such statistical techniques in modeling risk; however, they are still useful, provided we acknowledge these limitations and ensure that those who make decisions based on them are made aware of them. The dangers of modeling risks using statistical methods are abundant. Perhaps a recent one that has had the largest impacts on the greatest number of people is the stock market crash of 2008.

Back in 2000, a very clever mathematician named David Li introduced the Gaussian copula function to the world of quantitative finance. The formula was initially conceived as a quick way of assessing the financial risk of investments. It essentially simplified what was otherwise an infinite set of financial interactions into a rudimentary correlation of market prices based on credit default swaps (CDSs). In fact, Li's equation was what fueled an explosion in the use of credit default swaps and collateralized debt obligations (CDOs), which were instrumental in the spectacular crash of 2008. The notional value of derivative instruments exploded from $920 billion to $62 trillion between 2001 and 2007. Ironically, the inventor of the Gaussian copula function repeatedly tried, unsuccessfully, to alert Wall Street financial houses of the limitations of the equation—but Wall Street wouldn't listen.[6]

The lesson to be learned here is that quantitative risk analysis will always be limited by the assumptions on which the calculations are based. Financial markets are far too complex to include all data, and any quantitative model is therefore only as good as the information and algorithms used, which can never match the real world. A model, whether it is a risk model or a model of an airplane, will always be an imperfect facsimile of reality to a certain extent.

The failures of risk management are not limited to the misuse of quantitative modeling techniques. Drawing upon another example, let us consider the recent economic and ecological disaster of the Deepwater Horizon oil spill of 2010 in the Gulf of Mexico.

This is an excellent example of an event that was "not supposed to happen." In fact, the insurance industry considered the probability of such an event occurring as zero.[7] The event was a true Black Swan. The probability of the explosion happening was considered to be statistically insignificant given the various safeguards and procedures that were in place to prevent such occurrences. Therefore, it was considered by those in the industry to be extremely unlikely, and essentially "off the curve."

The initial explosion killed 11 people and resulted in the release of 1.5 million to 2.5 million gallons of crude oil per day into the Gulf of Mexico for a period of 86 days, impacting vital fisheries and decimating tourism revenues in the Gulf.[8] Further, the long-term damage to the environment is unknown but assumed to be severe. The ramifications of this catastrophe will have far-reaching impacts on government energy and environmental policy and will undoubtedly shape individual attitudes and behavior about energy and the environment for years to come.

Sadly, as befits the pattern of Black Swans, the event is already being explained away as an anomaly or "fluke" rather than as a risk that is fundamental to the nature of offshore oil production.[9] This means that it is likely that history will repeat itself as time passes and our collective memories fade. Indeed, this event was not the first such oil spill of its kind in history.

The Deepwater Horizon catastrophe should not have happened, and yet it did. It is likely that years will pass before all of the relevant factors that contributed to this disaster are brought to light. Nonetheless, this was a "known" risk and procedures had indeed been put into place to prevent it. The management of this risk failed as these preventative measures were not followed. Further, no contingency planning had been done, which meant that there was nothing to immediately fall back on when the blowout occurred. If there had, it is likely that the severity of the impact would have been far less. This is not only a failure of British Petroleum and the oil industry, but also a failure of federal regulatory agencies for not requiring such plans to be in place. So, in this case, we have a systemic failure of risk management from the standpoint of response planning and the monitoring and control of risk.

Risk management requires constant effort at every phase to maintain its efficacy. Even the most thorough and well-executed risk management plan is not infallible. We cannot predict the future, nor should we pretend to. However, if we approach the management of risk openly, honestly, and with great care and effort, we can minimize the effect of uncertainty on our projects.

OBJECTIVE OF THIS BOOK

The purpose of this book is to provide project managers, design professionals, estimators, schedulers, and contractors a systematic approach to manage project risk, specifically as they relate to cost and schedule.

An appropriate metaphor for this process is that of a bridge. Beneath the bridge flows a river of uncertainty. One side of the bridge is supported by the technique and application of risk assessment, which focuses on the identification and analysis of risks. The other side is supported by the remaining processes of risk management. The first span of the bridge is the risk-based estimate, which provides a means for us to get halfway across the bridge. The remaining span is supported by the piers of risk response planning, monitoring, and control, as illustrated on Cartoon 1.1. The point is that all of these actions are essential in order to manage project risk.

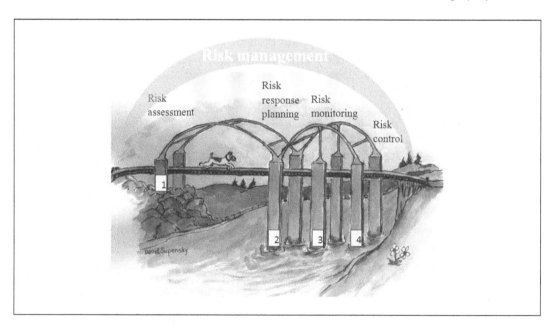

Cartoon 1.1   The Bridge of Risk Management

After reading this book and becoming fluent with the techniques and software applications that are presented, the reader should be able to:

- Understand that risk management is an essential component of project management that includes all of the necessary steps from risk identification through risk monitoring and control.

- Understand and use the process of integrated, quantitative cost and schedule risk analysis which we call the risk-based estimate (RBE).

- Avoid the common mistakes that users often make when risk-based estimating is employed.

- Understand and avoid the danger of professional sophistication as it applies to the risk-based estimate.

- Separate and define the two major components of risk-based estimates:
  - Base estimate
  - Risk events

- Develop a clear definition of base cost and schedule uncertainty by employing the probability box approach to consider:
  - Base variability
  - Market conditions

- Understand the different distributions and how they are best applied to risk-based estimates.

- Assess cost and schedule risks by considering:
  - Significant risks
  - The interrelationship of risk dependencies and correlations that form a project's "risk mesh"

- Assess the statistical impact of risks and base uncertainties on a project in terms of schedule and cost using mathematical models, specifically the Monte Carlo method, by:
  - Employing the self-modeling risk-based estimate spreadsheet
  - Understanding the true effect of a project's risk mesh, which is composed of:
    - risk dependencies
    - risk correlations
    - schedule sequence of risk events

- Interpret the true meaning of the results of risk analysis in terms of:
  - The range and shape of estimated cost and schedule histograms and cumulative distribution functions
  - The candidates for risk response planning

- Develop risk management strategies to minimize threats and maximize opportunities.

- Develop detailed action plans to implement the risk response strategies.
- Monitor and control risks throughout the life of the project.
- Improve project outcomes by applying the theory and techniques presented herein in a timely and conscientious manner.

ENDNOTES

1. Nassim Nichloas Taleb, "The Black Swan: The Impact of the Highly Improbable." *New York Times*, April 22, 2007.
2. *Project Management Body of Knowledge*, 4th ed., p. 273, Project Management Institute, December 31, 2008.
3. "Committee Draft of ISO 31000 Risk Management." International Organization for Standardization, 2009.
4. B. Flyvbjerg, M. Holm, and S. Buhl, "Underestimating Costs in Public Works Projects," *Journal of the American Planning Association*, Vol. 68, No. 3, Summer 2002, Chicago, IL.
5. "Managing Mega Projects." *World Finance*, December 17, 2009.
6. Felix Salmon, "Recipe for Disaster: The Formula that Killed Wall Street." *Wired Magazine*, February 23, 2009.
7. Byron King, "Too Much Debt, Not Enough Oil." *The Daily Reckoning*, July 27, 2010.
8. "BP: New Cap Has Stopped Flow of Gulf Oil," National Public Radio, www.npr.org, July 15, 2010.
9. "Gulf Spill Is 'Black Swan' Event: Industry Insider." CNBC, www.cnbc.com, June 9, 2010.

CHAPTER 2

PROJECT COST AND SCHEDULE ESTIMATES

INTRODUCTION

This chapter will discuss the need for and the importance of a good cost and schedule estimate when performing risk-based estimating. You will notice in the first sentence that terms *accurate* or *exact* were not used with regard to the word *estimate*. Estimates are completed at various stages of a project. The stages can be planning, scoping, design, letting, or construction. The estimating approach that is used on a project must align with the information available to the estimator at the time. As long as the estimate is as complete as possible in relation to the stage of the project, this will assist in providing reliable results when using this estimate in the risk-based estimating process.

Estimators should be shielded from pressures to keep estimates and schedules within programmed or desired amounts based on funding availability. Estimators should be free to establish what they consider to be a reasonable estimate and schedule based on the scope and timing of the project and the anticipated bidding conditions.

Historically, estimates developed at the later stages of a project tend to be more reflective of actual costs than those performed earlier in the delivery process. Indeed, estimates that are completed in the planning, scoping, or early design phases of a project, where there is very little known about the project, have been as much as 100 to 400 percent lower than what the actual cost of the completed project ended up being. This typically results in financial hardship for the owner of the project, whether it is private or public. Also, this has resulted in a loss of confidence by the public in the ability to estimate and deliver a project for the projected cost.

Deterministic estimates are typically expressed as a single number. Even though these deterministic estimates usually include contingencies, these are often inadequate and fail to cover the unexpected changes that projects experience. These changes are mostly a result of poorly defined scope, undocumented and endorsed schedule and estimate, estimate optimism, and risks that have come to fruition. The following are a few terms that are commonly used within the estimating profession and also throughout this book.

Cost Estimate—A prediction of quantities, cost, and/or price of resources required by the scope of an asset investment option, activity, or project. As a prediction, an estimate must address risks and uncertainties. Estimates are used primarily as inputs for budgeting; cost or value management; decision making in business, asset, and project planning; or for project cost and schedule control processes. Cost estimates are determined using experience and calculating and forecasting the future cost of resources, methods, and management within a scheduled time frame.[1]

Schedule Estimate, or simply "schedule"—Identifies "a plan of work to be performed, showing the order in which tasks are to be carried out and the amount of time allocated to each of them."[2]

It is important to emphasize that one cannot have a complete project estimate without having both a cost and schedule estimate. The two are mutually dependent, especially when considering the time value of money.

Base Cost Estimate—*Base cost* is defined as the cost which can reasonably be expected if the project materializes as planned. The base cost estimate must be unbiased and neutral—it should not be optimistic or conservative. The base cost estimate can include such things as miscellaneous item allowances and other adjustment factors such as costs related to fluctuations in commodities such as oil and steel. Caution must be exercised when including such items so that they are not captured as market condition risks, which would likely result in double counting. It is very important to clearly identify whether or not the estimate is in current year dollars or year of expenditure dollars.

Base Variability—Represents quantity and price variations related to the estimated base cost. Base variability is inherent in the base estimate. Base variability is always present and is not caused by risk events. It is captured as a modest symmetric range about the estimated value; that is, of the form: base value +/− X% (typically from +/− 5 to +/−10 percent depending on the level of project development and complexity). Base variability represents one form of uncertainty known as *epistemic risk*, meaning it is reducible as additional information becomes available. This concept will be explored in greater detail in subsequent chapters.

Example of Base Variability

When we decide to fill the gas tank in our car we have a general idea of what we will pay per gallon—but do not know for sure. Until we actually get to the gas station and make the purchase, there is some uncertainty about the cost per gallon. Similarly, if the gas tank is rated at 20 gallons and the gas gauge indicates half a tank, that informs us the approximate amount of gas *needed*—half a tank indicates about 10 gallons. When we fill it up, the actual amount will likely fall somewhere between 9 and 11 gallons, not precisely at 10.0 gallons.

Cost-Based Estimate—A method to estimate the bid cost for items of work based on estimating the cost of each component (labor, materials, equipment, including contractor and subcontractor markups) to complete the work and then adding a reasonable amount for a contractor's overhead and profit.

Historical Bid-Based Estimate—This type of estimate tends to be a straightforward count or measure of units of items multiplied by unit costs. These unit costs are developed from historical project bids and may be modified to reflect project-specific conditions. This is the most common type of estimating. These costs need to sometimes be adjusted for inflation, location, and size of project, time of year, and other factors.

Parametric Estimate—A method to estimate the cost of a project or a part of a project based on one or more project parameters. Historical cost information, usually in the form of bid data, is used to define the cost of a typical transportation facility segment, such as cost per lane mile, cost per interchange, or cost per square foot. Historical percentages can be used to estimate project segments based on major project parameters. These methods are often used in early estimating, such as planning and scoping estimates.

Risk-Based Estimate (RBE)—Involves simple or complex analysis based on inferred and probabilistic relationships between cost, schedule, and events related to the project. It uses a variety of techniques, including historical data, cost-based estimating, and the best judgment of subject matter experts for given types of work, to develop the base cost (the cost of the project if all goes as planned). Risk elements (opportunities or threats) are then defined and applied to the base cost through quantitative modeling (i.e., Monte Carlo method) to provide a probable range for both project cost and schedule. This will be discussed in great detail throughout this book.

Engineer's Estimate—Typically the final estimate prior to request for funding approval, letting, or advertising of the project.

Construction Engineering (CE)—The project management effort (budget/cost) of taking a project from contract execution through construction and project completion. In early estimates this is typically expressed as a percentage of the construction cost, but later in the project development, this cost should be taken from your resource-loaded construction schedule.

Construction Change Order Contingency—A markup applied to the base cost to account for uncertainties in quantities, unit costs, minor changes to work elements, or other project requirements during construction.

Miscellaneous Item Allowance—Additional resources included in an estimate to cover the cost of known but undefined requirements for an activity or work item. Allowances are part of the base cost. These should be significant in the early phases of a project and should be diminished in the later phases of a project.

Preliminary Engineering (PE)—The effort (budget/cost) of developing a project through the planning, scoping, and design phases. Planning and scoping typically have separate budgets but are encompassed under design or preliminary engineering. The terms *design* or *design phase* are sometimes used interchangeably with PE.

Mobilization—Typically calculated as a percentage of the total of the construction cost estimate, mobilization is included in a project estimate to cover a contractor's preconstruction expenses and the cost of preparatory work and operations (such as moving equipment onsite and staging).

COST ESTIMATING METHODOLOGY

Estimating methodologies fall into one of four categories: parametric, historical bid-based, cost-based, and risk-based. These categories encompass scores of individual techniques and tools to aid the estimator in preparing cost estimates. It is important to realize that any combination of the methods will likely be found in any given estimate. Too often, designers feel that if they use one type of estimating method for a portion of the project, they need to use this for the entire project. Using a combination of the methods will improve the efficiency and accuracy of preparing an estimate.

The risk-based estimating process requires a base estimate, which could encompass a mixture of the parametric, historical bid-based, and cost-based methods for its development.

This book will not attempt to explain how to estimate as there is a large amount of published material on this subject already; however, we will attempt to explain what is needed for the base estimate that is used in the risk-based estimating (RBE) process. Also included are tips and lessons learned that the authors have accumulated over several years of performing risk-based estimating.

COST ESTIMATING PROCESS

All projects benefit from following a thoughtful and deliberate process in developing project cost estimates. In order to provide a typical example, the process presented in Figure 2.1 describes how an organization may develop its project cost estimates. It is applied to all levels of project delivery, beginning with the planning (conceptual) level and ending with the final estimate.

Each level of estimate may require different estimating inputs, methods, techniques, and tools. Also, the process is scalable to the level of estimate being prepared. It should be noted that the figure presented here includes an activity called "Determine Risks and Set Contingency." This activity should not be confused with risk-based estimating, which is discussed in great detail in Chapters 4 and 6. This activity is related to setting a "gross" placeholder that is often used in deterministic estimates to account for unknowns. In later chapters, when using the RBE process, these gross contingencies will be replaced.

The task of cost estimating, by its very nature, requires the application of prudent judgment to the completion of the task. Documentation of the use of judgment and experience is very important. Too often the task of preparing an estimate is handed off to the least experienced person in the organization and put off until the very end of the project.

The cost estimating process described in this book includes the following steps:

- Determine basis of estimate.
- Prepare estimate.
- Review estimate.
- Determine risks and set contingency.
- Determine estimate communication approach.
- Conduct independent review and obtain management endorsement.

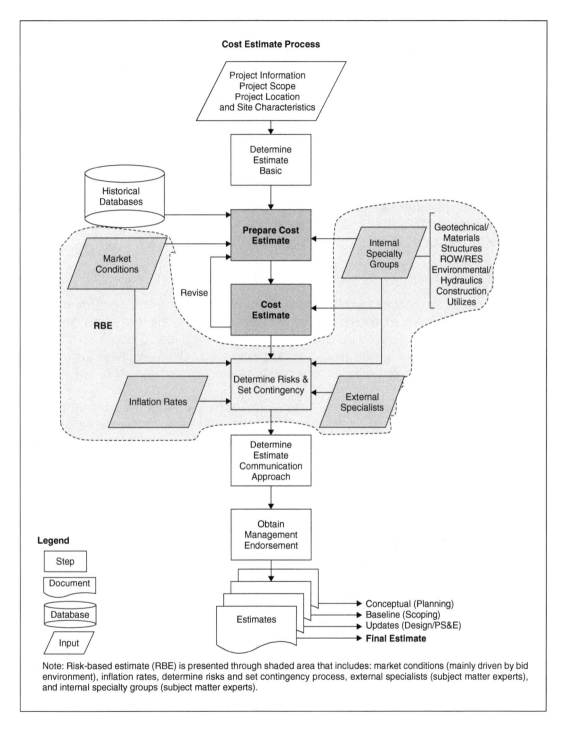

Figure 2.1   Cost Estimating Process

Determine Basis of Estimate

Getting started is one of the most important steps in the cost estimating process. The phrase "one must walk before running" is all too true in the estimating process. Without a good, clear understanding of the project, an estimator often neglects to capture the project-specific or site location effects that will influence the prices or quantities used for developing an estimate.

The basis of estimate (BOE) activity focuses on obtaining project information, including all previously developed project scope and schedule details and data, from which a project cost estimate can be prepared. The level of scope detail varies depending on the project phase, type, and complexity, but would include the design criteria, all assumptions, and pertinent scope details.

The estimate basis should be clearly documented and forms the beginning of the estimate file that should be prepared for each estimate. Each of the following steps will add information to this file, with the end result being a complete traceable history for each estimate. This documentation is essential for use in the risk-based estimating process.

Prepare Estimate

This activity covers the development of estimated costs for all components of a project, excluding future escalation. These components may be estimated using different techniques depending on the level of scope definition and the size and complexity of the project. The number and detail of components estimated may vary depending on the project development phase. For example, in the scoping phase, the cost estimate covers preliminary engineering, right of way (i.e., real estate), and construction.

As the design progresses and more details are known, pieces of the estimate become more detailed. Key inputs to this activity include project scope details, historical databases and other cost databases, knowledge of market conditions, and use of inflation rates. The use of specialty groups should be included to provide estimate information when preparing estimates. As an example of this, you wouldn't have a plumber give you an estimate on fixing your car, or conversely, you wouldn't have your auto mechanic give you a plumbing estimate.

The estimate should also be based upon, and include as an attachment for reference, the associated schedule for all remaining project activities. For conceptual level estimates the schedule will be cursory and very broad in its coverage. However, as a minimum it should include major milestones. The conceptual level schedule may only include a few activities, but should begin with the development of the project, and include the right of way, design, and construction phases. This schedule will be used when the risk-based estimating process is done for the project. The schedule will assist with the development of the flowchart of the project, as discussed in Chapter 3.

Another required component of the estimate step is the preparation of a basis of estimate document that describes the project in words and includes underlying assumptions, cautionary notes, and exclusions. This documentation should tell a story about the project and what went into the estimate. This documentation will be used continuously throughout the development of a project and also during the risk-based estimating process to give the subject matter experts an understanding of what went into the estimate.

Review Estimate

This activity is necessary to ensure that (1) assumptions and basis are appropriate for the project; (2) the cost estimate is an accurate reflection of the project's scope of work; (3) scope, schedule, and cost items are calculated properly and required components are not missing or double counted; and (4) historical data, the cost-based estimate data, or other data that were used reasonably reflects project scope and site conditions. Internal specialty groups and/or subject matter experts (SMEs) must participate in reviewing the estimate. This can also be considered the validation of the estimate.

Determine Risks and Set Contingency

This step is likely one of the most difficult and controversial steps of preparing an estimate. In traditional cost estimating, it is customary to assign a percent factor to address project uncertainties related to design and construction. This process is typically regarded as more art than science, and is largely based upon estimating judgment and experience. Contingencies typically diminish as a project matures. The determination of risks under this traditional estimating approach is gross in nature and not calibrated to speak to individual risks but rather characterize the general level of project uncertainties.

Determine Estimate Communication Approach

Cost estimate data is communicated to both internal and external stakeholders in both the public and private industries. The communication approach determines what estimate information should be communicated, who should receive this information, how the information should be communicated, and when the information should be communicated. Cost estimate information should be included when the communication plan is developed as part of the project management process. Often the words are as important as the numbers. The BOE document can be used effectively as a communication tool to convey key information about the project to others.

The number that is given early in the development of a project is the number everyone remembers. Therefore, communicating estimates in ranges is strongly encouraged in the early planning, scoping, and design stages of a project. As discussed in the following chapters, the risk-based estimating process produces a range of numbers that can be used to communicate the cost of the project.

Conduct Independent Review and Obtain Management Endorsement

Estimates are key products of the project management process and are fundamental documents upon which key management decisions are based. Early estimates also establish the baseline for the project for which the customers often measure the success or failure of a project. Given their importance, all estimates should receive an independent review and then be reconciled and revised as needed to respond to the independent reviewer's comments.

Once independent review comments have been satisfactorily addressed, estimates should be presented to management staff for approval.

Management approval of estimates developed for initial budgeting or baseline definition is a defined step in the project management process. Revised estimates are typically developed if project requirements change, or as design is developed; these should also be reviewed by management staff, revised as necessary to reflect management comments, and then approved. It is very important to document the changes and the reason for a change so that this documentation tells a story about the progress of the estimate.

COST ESTIMATING DATA

An estimator must be able to identify the needed completeness of the data relative to the development stage of the project. The estimate is intended to serve as a snapshot in time for the project and decisions need to be made about the data that is needed. Too often estimators continuously put off producing an estimate with the thought being, "If only I can have another day or two, the accuracy will be much better." This line of thinking will continue until decisions are made to take the information and use it. This is why documentation is so important, which will be covered later in this chapter.

COST ESTIMATING AND PROJECT DEVELOPMENT LEVEL

There are four main phases, or levels, of project development:
1. Planning
2. Scoping
3. Design
4. Final estimate or bidding/letting

The estimate for each level of project development has a specific purpose, methodology, and expected level of accuracy. Figure 2.2 summarizes the relationship that exists between project development levels and the estimate's expected level of accuracy. Note the inverse relationship between the project development level and the expected accuracy range. Some of the typical causes of project cost uncertainty are a lack of scope definition, multiple alternatives, and a lack of information about factors outside the project premises such as: real estate, community, cultural, and environmental. As the project progresses, more data is available and the expected accuracy range narrows.

Figure 2.2 displays the situation of a project that may end up to be constructed for $100 million. By choosing this number the interim cost estimates could easily be related to relative percentiles. The range recommended has a wide variation as the magnitude and gradient for the different levels of design mature.

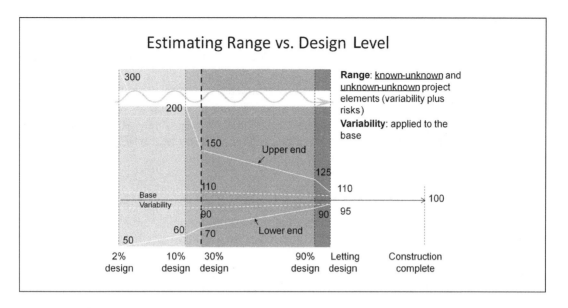

Figure 2.2   Cost Estimating Range and Design Maturity

The range asymmetry is quite noticeable for all levels of project design maturity. The upper end is farther away from the baseline estimate than the lower end. At the same time, the gradient of the upper end is higher than the gradient of the lower end. Once the project's estimators learn more about the project-specific elements, the upper-end cost moved rapidly toward the baseline.

The fact that the upper end of the cost estimate is more dynamic than the lower end suggests that throughout the life of the project there are economic, political, and social forces that put pressure on reducing the project cost estimate and perhaps the project schedule. While these findings are not surprising, it is important to recognize this tendency and watch for it in the cost estimate review process.

The project design maturity may be divided into two main phases:

- Planning, including scoping, when the projects may have a vague definition that might justify the wide range of the estimate. It is assumed that a project may have a clear definition at approximately the 30 percent design level.

- Design phase, when the range of the estimate is defined by uncertainties in the cost of materials and labor plus "environmental conditions."

The projects may acquire a better cost baseline value after 5 to 10 percent design when the project definition starts to be better defined. The scoping phase (10 to 30 percent) focuses on establishing a clear project definition, and at the end of this phase, many scope uncertainties are clarified. During this phase the cost and schedule baseline are better developed, and this allows establishing the project budget.

The planning phase may have the predominant focus on scope management and the design phase may focus on risk management.

The design phase (30 to 100 percent) deals with uncertainty in cost of material and labor plus environmental conditions such as: market condition, inside or outside pressures, and events (risks) that may change significantly the project cost and schedule. These variables generate an uncertainty in the project cost and schedule that must be captured by the RBE process.

Planning

The planning level estimate is used to estimate funding needs for long-range planning and to prioritize needs. These estimates are typically prepared with little detail to the project definition.

Parametric estimating techniques are often used for planning estimates. Lane mile and square foot are two types of parametric estimating techniques. Historical bid prices and historical percentages can be used to generate costs for these parameters. Analogous project estimating is another approach that can be used. Commercial estimating programs are available to assist in parametric estimating, especially for projects that have little or no historical data available.

Miscellaneous item allowance in design at this level of estimate typically ranges from 0 to 50 percent, and ranges even higher on nonstandard projects (see Table 2.1). This includes rounding costs (and quantities) to an appropriate significant figure.

Risks should be identified, and a risk management plan developed to be included in the estimate notebook for future reference.

When using analogous project estimating, the chosen historical project must be truly analogous. Finding an appropriate project or projects and determining the similarities and differences between the historical projects and the current project can take significant time and effort. Project data from older projects is less reliable due to variations in prices, standards, construction technology, and work methods. The analogous method is best used as a tool to determine broad price ranges for simple, straightforward projects or as a check to verify estimates prepared using another method.

Some concerns that estimators should be aware of are:

- Due to the lack of scope definition or preliminary design, care should be taken to properly communicate with project stakeholders regarding the range of possible cost and schedule changes as the project becomes more defined.
- Given the large-scale assumptions inherent in parametric estimating methods, the estimator must document all assumptions clearly.
- Estimators should guard against false precision; that is assuming a level of precision that is not inherent to this level of estimate. Although a properly developed estimate will include well documented assumptions, many of the details that impact project cost are not defined at the time a planning level estimate is done.
- Keep the estimate current as the project waits to move on to scoping.

TABLE 2.1   Cost Estimating Markups Summary

Cost Estimating Elements	Planning	Scoping	Design	Final /Letting
Mobilization	Typically a%	Typically a%	Typically a%	Typically a%
Sales Tax	Site-specific	Site-specific	Site-specific	Site-specific
Preliminary Engineering	Typically a%	Typically a%	PM's work plan + Actual to date	PM's work plan + Actual to date
Miscellaneous Item Allowance in Design[1]	30% to 50%	20% to 30%	10% to 20%	0% (all items should be defined)
Contingency	Typically a%	Typically a%	Typically a%	Typically a%
Construction Engineering	Typically a%	Typically a%	PM's work plan	PM's work plan

[1]Miscellaneous Item Allowance in Design accounts for lack of scope definition and those items too small to be identified at that stage of the project. This allowance is eliminated entirely in final estimates as the scope will then be fixed and all estimate items should be identified.

Scoping

A scoping level estimate is typically used to set the baseline cost for the project and to program the project. The scoping estimate is important because it is the baseline used to set the budget and all future estimates will be compared against it. Clearly document assumptions and scope definitions in the basis of estimate document so that all future changes can be accurately compared to this estimate.

Historical bid-based, cost-based, parametric, and risk-based are some of the techniques used while preparing scoping estimates. The estimator will be able to determine approximate quantities for items such as foundations, exterior walls, concrete flat slabs, asphalt pavements, excavation, and so forth. For such quantifiable items, historical bid-based or cost-based estimating methodologies should be used for pricing. Other items not yet quantified should be estimated parametrically or through the use of historical percentages.

Miscellaneous item allowance in design at this level of design definition typically ranges from 20 to 30 percent, and ranges even higher on nonstandard projects (see Table 2.1). This includes rounding costs (and quantities) to an appropriate number of digits.

Risks should be identified, and a risk management plan developed to be included in the estimate notebook for future reference.

Some concerns that estimators should be aware of are:

- Create/update the basis of estimate document. All changes, assumptions, and data origins should be clearly documented. This is particularly important because any future estimates will be compared with this one to justify changes in the cost of the project.

- Estimators should guard against false precision; that is, assuming a level of precision that is not inherent to this level of estimate. Although a properly developed estimate will include well documented assumptions, many of the details that impact project cost are not defined at the time a scoping level estimate is done.

- It is important to choose the correct unit costs for major items and then correctly inflate those costs to current dollars.

- Use sound risk identification and quantification practices to ensure that major risks to the project are identified and documented.

Design

Estimates prepared at the various design levels are used to track changes in the estimated cost to complete the project in relation to the current budget. Each time the estimate is updated, the cost estimate process detailed in Figure 2.1 should be followed. The current project budget and schedule should be compared to the new estimate. Clearly document each of these updates in relation to the previous estimate and include the documentation in the estimate file. This documentation will assist the estimator of future estimates in telling the story on why the project costs changed.

Historical bid-based, cost-based, and/or risk-based are some of the techniques used while preparing design estimates. As design definition advances, design engineers and estimators are better able to determine project work items and their associated quantities and unit prices. Historical bid-based methodologies are typically used for items of work for which historical data is available. Cost-based estimating methodologies can be used for those items with little or no bid history, or for major items of work that are project cost drivers. Key resources are suppliers and other individuals knowledgeable about current prices for the subject items, typical construction methodology and production rates, and equipment used. The estimator should contact these resources to develop basic cost data for materials, labor, and equipment.

If miscellaneous item allowances were used during the scoping phase, these should be reviewed and updated (usually reduced) as the level of development of the project increases.

Review risks identified earlier in the project development process and update the risk management plan to reflect the current design level and risks.

Some concerns that estimators should be aware of are:

- As with the scoping level estimate, estimators should guard against false precision—thinking they know more about a project than they do. Significant project definition continues to be developed until the project is ready for advertisement. Use appropriate item allowances and ranges for estimates.

- If cost-based estimating techniques are used, pay special attention to documenting all of the assumptions that are made in the development of unit prices such as the crew size, crew makeup, production rates, and equipment mix and type. The costs assumed for contractor overhead and profit as well as for subcontractor work should also be clearly documented. It is important to remember that these decisions may not reflect the decisions of the individual contractors that will bid the job, thus introducing elements of risk into the estimate.

Final Estimate or Bidding/Letting Phase

The engineer's estimate is prepared for the final contract review in preparation for advertisement or letting and is used to obligate construction funds and to evaluate contractors' bids.

Historical bid-based, cost-based, and risk-based are some of the techniques used while preparing final estimates. The project has matured to a point where design engineers and estimators are able to specify all items of work that will be required for the project and can accurately estimate quantities and unit prices. This level of project estimate has the advantage of detailed understanding of project scope and conditions. If the estimators are from outside the project team, they should take special care to understand the details of the project, including performing a detailed review of the plans and specifications. Clearly document the development of and adjustments to line item quantities and prices. This is critical for both the review of the estimate and the review of bids prior to award. This data should be clearly defined and identified in the estimate file. Historical bid-based methodologies should be used for most items of work where historical data is available. Cost-based estimating methodologies can be used for those items with little or no bid history, or to check major items of work that significantly impact on the total project cost.

Miscellaneous item allowances should not be included in an estimate at this level. All quantities and items should be known.

Some concerns that estimators should be aware of are:

- Reviews of these types of estimates should be extensive and detailed and should include final independent QA/QC checks of calculations, prices, and assumptions. The basis of estimate and overall estimate documentation package should be carefully reviewed to make sure they are complete, accurate, and easily understood, and that all figures, from detailed backup to summary levels, are traceable.

- Major quantities and cost drivers should be carefully checked to assure that they have been properly calculated (proper conversion factors have been used and allowances applied to neat line quantities, if applicable).

- Contract special provisions should be carefully reviewed and cost and schedule impacts incorporated into the engineer's estimate.

ESTIMATE DOCUMENTATION

Documentation is a key element in good estimating practice. The estimate file should be a well-organized, easy-to-follow history from the first estimate at the beginning of the planning phase through preparation of the final estimate. The BOE document, described in this section, contains recommended organization, topics, and format. Each estimate should track changes from the previous estimate, updating the scope, assumptions, quantity and price calculations, and risks from the previous estimate. At each update the differences between the previous estimate and the current estimate should be highlighted. This contributes to transparency and accountability in estimating and promotes consistency between estimates.

Clear documentation is particularly important as the project passes from one group to another, or as team members change. The project estimate file should follow the project through the various stages so that each new estimate can be easily tied to the previous one.

Several techniques can be employed to ensure clear documentation. It is recommended that estimating be specifically scheduled in the project management plan for each phase of the project. This ensures that adequate time and resources are allotted for performing the estimate. A specific schedule should be developed for each estimate that includes the steps outlined in Figure 2.1. As part of the estimate review process, someone external to the project team should perform a review of the estimate file. This external review will help ensure that the estimator has clearly recorded the assumptions and decisions made in the estimating process. One form of estimate documentation that the authors have used is the basis of estimate. This has worked well and is described below.

BASIS OF ESTIMATE

The basis of estimate (BOE) is characterized as a document that defines the scope of the project, and ultimately becomes the basis for change management. When the BOE is prepared correctly, it can be used to understand and assess the estimate, independent of any other supporting documentation. A well-written BOE achieves these goals by clearly and concisely stating the purpose of the prepared estimate (i.e., cost study, project options, benefit/cost study, funding, and so forth), the project scope, pricing basis, allowances, assumptions, exclusions, cost threats and opportunities, and any deviations from standard practices. The BOE is a documented record of pertinent communications that have occurred and agreements that have been made between the estimator and other project stakeholders. The authors have found that the use of a BOE is an excellent tool when validating the estimate in a risk-based estimating workshop.

A well-prepared basis of estimate will:
- Document the overall project scope
- Document the items that are excluded from the project scope
- Document the key project assumptions

- Communicate the estimator's knowledge of the project by demonstrating an understanding of scope and schedule as it relates to cost
- Identify potential risks (threats and opportunities)
- Provide a record of key communications made during estimate preparation
- Provide a record of all documents used to prepare the estimate
- Provide the historical relationships between estimates throughout the project lifecycle
- Facilitate the review and validation of the cost estimate

Points of significance when preparing a BOE include:

- Factually complete, yet concise
- Ability to support all facts and findings
- Identify estimating team members and their roles (including specialty groups)
- Describe the tools, techniques, estimating methodology, and data used to develop the cost estimate
- Identify other projects that were referenced or benchmarked during estimate preparation
- Develop and update the cost estimate and the BOE concurrently
- The BOE establishes the context of the estimate, and supports review and validation.

Additional Tasks for the Estimator

At the conclusion of completing an estimate, the estimator should assemble the following items if the project will have a risk analysis performed on it:

- Ensure that the estimate and schedule have been reviewed and are current.
- Document the risks that were identified during the preparation of the estimate.
- Identify the approximate base cost variability that should be applied to the base cost estimate. This will be reviewed and verified during the risk analysis.
- Ensure that the BOE or other form of estimate documentation is complete.

CONCLUSION

The deterministic estimate described in this chapter forms the foundation for the risk-based estimate. It is therefore essential that this starting point follow good practices and be well documented. The quality of the RBE process will largely be predicated on the quality of the estimate and its basis.

The basis of estimate must provide a sufficient level of detail and clear information about all of the assumptions and constraints that help shape the estimate. This information forms the backbone of the RBE process and will serve to inform the RBE workshop participants about the project as it relates to cost.

Using the steps described in this chapter, the estimator will be well-prepared to present the estimate and schedule to the risk analysis team for validation as described in Chapter 3.

ENDNOTES

1. Copyright 2007, AACE International, Inc., AACE International Recommended Practices, Number 10S-90.
2. *Encarta* online dictionary, s.v. "schedule," accessed March 2010, http://encarta.msn.com/encnet/features/dictionary/DictionaryResults.aspx?lextype=3&search=schedule.

THE RISK-BASED ESTIMATE

RBE—THE PROCESS

The risk-based estimate (RBE), known as cost risk analysis, is an important part of project management as well as project cost and schedule estimating. Chapter 2 focused on deterministic estimates as a method of developing cost and schedule estimates, yet this method lacks the capability to effectively deal with the multidimensional nature of uncertainty. If we imagine deterministic estimating as a simple axis, the estimated project cost is represented by a single point moving along the axis (Figure 3.1).

Figure 3.2 supplements this representation with a horizontal axis that represents cost-only risk events. In other words, Figure 3.2 illustrates the situation when the analysis assumes that changes may occur only for the cost and the schedule is fixed. Later in the chapter, Figure 3.38 presents another angle of cost-only risk events. Each risk is represented by its probability of occurrence and its impact's range and shape. At the same time the axis of the deterministic estimate is enriched by capturing on it the range and shape of a cost estimate's deterministic value while considering the randomness of market conditions (market conditions will be defined in detail later in the chapter). The range and shape along the vertical axis represent the estimate's uncertainty (which will be explained in greater detail later in this chapter as well). Combining the data provided by these two axes allows for a two-dimensional expression of the project's cost estimate.

In other words, combining a deterministic estimate with its associated cost risk events provides a means to consider cost risk analysis. This new process generates a richer set of data that lends itself to proactive project risk management. Despite this enhancement the process is limited because it does not consider the effect of time.

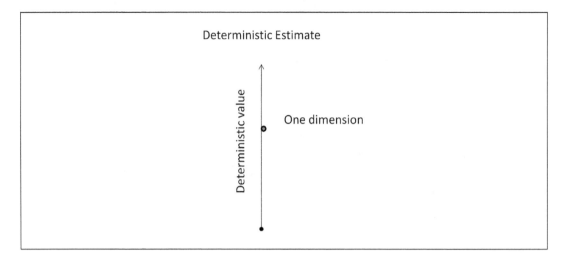

Figure 3.1   One Axis—Deterministic Estimate

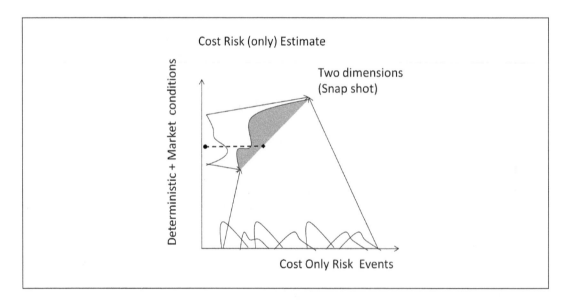

Figure 3.2   Two Axes—Cost-Only Risk Analysis

The next level of cost risk estimating considers the effect of time, including schedule risks. In this case, cost-only risk analysis is combined with schedule risks and the related effects of inflation. This integration of cost and schedule risks generates a much richer data set and provides for a far more robust vehicle to effectively manage a project's risks (Figure 3.3).

Figure 3.3 illustrates the three-dimensional complexities of integrated cost and schedule risk analysis. The schedule-only risk events axis induces a time component into the analysis. Usually

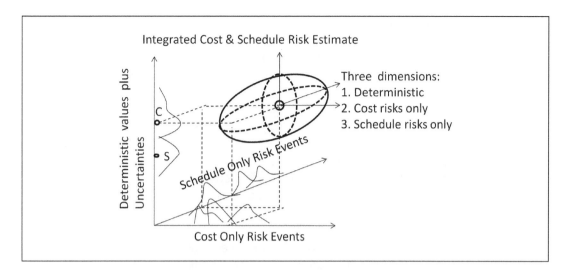

Figure 3.3 Three Axes—Integrated Cost and Schedule Risk Analysis

most projects develop data about cost and schedule; however, they are not always rigorously connected. Integrated cost and schedule risk analysis requires the development of an algorithm that may blend them together and present the combined results.

This chapter explores cost risk analysis as an integration of cost and schedule risk estimating. Terms such as *stochastic estimate* (SE), *risk-based estimate* (RBE), and *quantitative risk analysis* (QRA) are other common expressions used to describe the cost risk analysis process. The deterministic estimate, as presented in Chapter 2, should be used for determining cost estimates and providing the basis for the development of the risk-based estimate. As will be discussed later in this chapter, the deterministic estimate is fundamental in defining the "base cost and schedule" of the RBE.

Risk-Based Estimates — Advantages and Disadvantages

Chapter 2 focused on the deterministic estimate. The simplicity of this form of estimating ignores the effect of uncertainty and therefore has inherited a number of shortcomings. These include:

- It creates an expectation of certainty that in reality does not exist.
- It does not consider risk events that could change the estimate in a positive or a negative way.
- There is little or no opportunity to proactively manage risk.
- It allows for little control over a project's estimate.
- It is reactive and, in the majority of cases, any remediation that emerges from it is likely to be more costly than it should be.

■ It requires a sagacious estimator who has developed the ability to determine appropriate project contingencies for different situations. Such expertise is more akin to an art form and requires a deep knowledge of construction estimating coupled with years of experience. Not surprisingly, there is a dearth of "sagacious estimators" in the world today.

The development of a risk-based estimate is based on a structured approach that involves a collaborative team effort. Further, it requires training in risk elicitation and analysis. The process is quite similar to the *Successive Principle* or *Intelligent Cost Estimating*.[1] Developing an RBE is also more resource-intensive than producing a deterministic estimate for the same project. Despite the added complexity, effort, and training required, this method is reproducible and, ultimately, more reliable than traditional methods. The advantages of using the RBE approach include:

■ It minimizes the number of surprises during project development and delivery by providing for the identification and quantification of risk events. In other words, it increases chances of a successful delivery of the project.

■ It creates opportunities to study "what-if" scenarios using a rigorous and statistical approach.

■ It allows reasonable control over the project's estimate through project risk management.

■ It is a collaborative effort that improves project communication and transfer of information among the project team members, stakeholders, and the public.

■ Realistic contingency planning (risk reserve) is made possible since it considers the effect of positive and negative events that may affect the project.

In summary, RBE gives program and project managers a sharper and more realistic long-distance view of the prospects awaiting their projects.[2] This foresight will help project managers prioritize their efforts; focus resources more effectively, and take decisive action as necessary to manage cost, schedule, and risk.

Risk-Based Estimate — Keep It Simple . . .

Since 2002 the authors have analyzed and evaluated hundreds of estimates produced through a rigorous RBE process.[3] During this time, the RBE process underwent an iterative development cycle that took into account the number of variables, the type of distribution, and the role of market conditions to name just a few key issues.

The RBE requires intense efforts but is worthwhile, and the results extend beyond the final analysis as presented in previous paragraphs. We present a quote from Bob Stromberg, senior program manager, whose message the authors have heard repeatedly on various projects: "Even without having the results yet, this process is worth far more than the money we have spent on it just because of the communication between the subject matter experts (SME) and support offices."

Because RBE can require a significant investment, there is a tendency to increase its complexity to better justify the expenses. We call this tendency *professional sophistication* and it appears when a RBE includes too many activities, too many variables, or too many insignificant risk events, to name a few of the usual suspects. Professional sophistication is detrimental to cost and schedule risk analysis[4] and constitutes a significant source of error.

Rather than using the tired phrase, "Keep It Simple, Stupid," we will instead use the mantra "Keep It Simple, Smarty" (KISS). In other words, a smart professional will always maintain a balance between professional sophistication and simplicity.

In the next few pages, a number of case studies are presented where so-called professional sophistication led to failure of the initial analysis and, if not caught in time, could lead to a significant misrepresentation of the project's cost and schedule. At the same time, we will show the reasons why the KISS principle is critical for a robust and reliable cost and schedule risk analysis.

Professional Sophistication versus KISS — Rival Methodologies or Frameworks

Professional sophistication is best exemplified by examination of the two equations presented in Figure 3.4. Given today's computational technology, these equations could be written in an unending string of numbers extending from Seattle to Boston.

While these equations are, of course, mathematically correct (assuming they do not contain a typo, which is indeed a risk we must consider), they express a simple truth, which is presented in Figure 3.5. Of course, this figure is meaningless. Everyone knows that "$1 + 1 = 2$." The intent is to represent the KISS alternative of professional sophistication presented in Figure 3.4.

$$(\sin \alpha)^2 \sqrt{1 + (\cot \alpha)^2} + \log 10 = \sum_{1}^{\infty} \frac{1}{2}(\frac{1}{2})^n$$

$$(\cos^2 \alpha + \sin^2 \alpha)^{\sqrt{a^2+b^2}} + [\sum_{k=0}^{n} \binom{n}{k}x^k a^{n-k} (1 + x)^n]^{(\cos^{-1} x - \sec x)} =$$

$$\{\ln [\lim_{n \to \infty} \left(1 + \frac{1}{n}\right)^n] + (\sin \alpha)^2 \sqrt{1 + (\cot \alpha)^2}\}$$

Figure 3.4   Sophisticated Formulas

$$1+1=2$$

Figure 3.5   KISS Equivalent

Regarding the equations presented on Figure 3.4, we were careful enough to state that they are correct provided there are no typographical errors present. If typos are indeed involved (just one little error is needed), the equation is simply wrong. To make matters worse, identifying such an error is difficult, if not impractical. The crux of the problem is that a computational error will incorrectly inform the decision-making process. When this happens, things can go south quickly.

In the next few pages several case studies will be presented that show how professional sophistication can destroy the credibility of the data that it produces. The scenarios represented were actual situations that occurred and involved real experts in the field of risk analysis and cost estimating. The intent of these examples is not to cast doubt on the expertise of anyone but rather to illustrate that "typos" do happen and that when they do, it can be extremely difficult to identify and correct them. When quantitative risk analysis is more sophisticated than it should be the potential for creating a flawed outcome based on computational errors significantly increases.

Scenario 1 — $1 Billion Project

This case study involves a project that went through a combined risk-based estimate and value engineering (VE) study. The initial action involved developing an RBE followed by a VE study that was intended to focus on developing response strategies to address risks identified during the first week. The results were meant to include the risks quantified in the first week and the risk response strategies identified during the VE study.

The model developed by the risk consultant was sophisticated and included more than 100 activities with each activity being loaded with variables for base cost and base duration. More than 160 risks were elicited and quantified, including 10 major VE recommendations that were treated as risks (mainly as opportunities).

The draft report was presented to the project team for review, and Figure 3.6 represents one of the report's findings. Figure 3.6 shows that risk response strategies identified during the VE study led to significant reduction in cost. The shape of the curves suggests that they were produced by approximately $200 million in savings. This conclusion is backed up by the data presented in Table 3.1. The base cost represents the estimated cost of the project if the project is delivered as planned (i.e., no risk events occur).

Figure 3.6 suggests that the base cost may be significantly reduced when the cost reduction resulting from the VE alternatives are considered (the curves representing the effect of the VE alternatives are similar to the original curve but shift to the left by $200 million) but Table 3.1 shows no significant change to the base cost. This kind of situation is unusual in risk analysis but possible if the project is affected by opportunities of about $200 million that have near 100 percent probability of occurrence.

The next step of this examination considers the risk events in order to find the "big opportunities" that may have created the unusual result. From the draft report we have identified the "key VE alternatives" as presented in Table 3.2.

Table 3.2 shows that many of the key VE alternatives do not reduce the cost and in actuality increase the cost and confirm the variation of the base cost presented in Table 3.1.

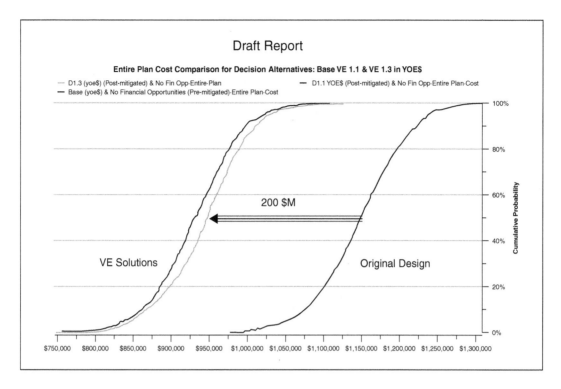

Figure 3.6   Draft Report Results (the cost is in $1,000)

TABLE 3.1   Cost Distribution (in millions)

Category	Original Design	VE 1.3	VE 1.1	VE 1.3 Savings	VE 1.1 Savings
Base cost	1,037	1,047	1,003	-10	34
Mean cost	1,149	944	931	205	218
1%	977	747	753	230	224
10%	1,075	866	862	209	213
20%	1,101	898	886	203	215
30%	1,121	919	903	202	218
40%	1,137	935	918	202	219
50%	1,150	948	931	202	219
60%	1,163	960	947	203	216
70%	1,179	973	961	206	218
80%	1,196	990	977	206	219
90%	1,222	1,010	997	212	225
99%	1,308	1,127	1,110	181	198

TABLE 3.2   Summary of Key VE Alternatives

VE Alternatives	Amount
Building relocation permanently	+23.6 $M
Building relocation temporarily	-15.8 $M
Hybrid wharf	-10.0 $M
Interim at grade intersection	+27.2 $M
Construct from land	24.2 $M
Shorten temporarily the JJ dock	-7.0 $M
Utilize Trinity area for Pacific Lines	-2.6 $M
Phase construction of Pacific Lines dock	-6.1 $M
Total additional cost	**+33.4 $M**

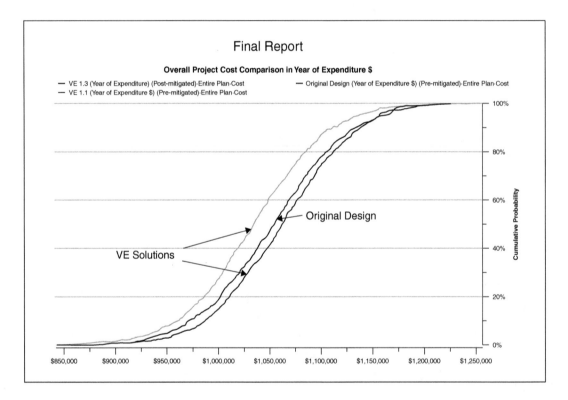

Figure 3.7   Corrected Results (cost is in $1,000)

"VE Alternative 1.3" has the base cost reduced by $10 million and "VE Alternative.1.1" has an increase in the base cost by $34 million. So the question was: "Where is the $200 million cost avoidance coming from?"

A careful reexamination of the model uncovered several "typos" and after they were fixed, the model provided the results presented in Figure 3.7.

Figure 3.7 looks different than Figure 3.6. The $200 million cost avoidance is no longer in the picture. The results intuitively made more sense to the project team as well, since the VE study was focused on reducing the project schedule even if it resulted in increased costs.

Scenario 2 — $90 Million Project

This example is simpler than the previous one. In this case, the model wasn't too "sophisticated" but still complicated enough for the modeler to get lost in it. The subject of the analysis was a project in a rural area that cost about $50 million in construction, $12 million in preconstruction activities, and about $10 million in real estate acquisition. The modeler had sent us the preliminary results and one of the tables intrigued us. Table 3.3 presents the results.

The reader may see immediately that something is wrong with the numbers presented in Table 3.3. The ROW (i.e., real estate) cost is enormous. The disproportion between construction cost and ROW cost is unprecedented. The ratio of ROW to construction is about 20:1, which is greater than one might expect even in the heart of Manhattan.

Upon reviewing this model, one of the authors responded with an email to the modeler: "Have you checked your results?" The answer came back a day later:

"I have checked my inputs and am satisfied the model is working correctly, however, there is a small discrepancy in my cost-loaded resources and estimate."

The model complexity made the modeler ignore the question and answer a different question. It is an example of losing track of the scope of the work. The authors believe that this situation is related to treating things with greater complexity than is needed—professional sophistication.

Common Scenarios

The last two scenarios present significant consequences related to mistakes rooted in professional sophistication. The scenarios may be rare occurrences; however, one should be aware that they do occur. The most dangerous ones are the cases presented in the next few examples because they are repeated and, in most cases, no one is ever alerted to them. These scenarios are the product of the professionals who create and develop sophisticated models. The mentality

TABLE 3.3 $90 Million Project Estimated Cost

Alternative 2b with Escalation	Total	PE	ROW	CONSTRUCTION
10%	$972,568,448	$11,746,794	$875,814,783	$58,226,961
50%	$1,016,503,702	$12,885,815	$918,097,390	$59,431,655
90%	$1,674,601,286	$13,746,095	$1,573,448,151	$60,767,961

"more is better" or "sophistication is expertise" drives many professionals away from the KISS principle.

Number of Variables and Their Significance

Perhaps the most important recommendation that can be made to any risk analyst is the following: Risk analysis should adopt a comprehensive approach rather than a fragmented one and it should be tailored to the quality of data available. The quality of a project's data depends on such things as the project's scope, level of design, database used, staff's experience, and so forth. For example, for a project at the 5 percent design level, the quality of data available doesn't allow for consideration of a "$1 million" risk to be included on a "$1 billion" project that is being analyzed. The KISS approach urges risk analysts to heed the following advice:

- The estimating detail should match the data available.
- Risk elicitation should engage the project as a whole and focus on a broad range of project areas such as geo-technical, structural, construction, environmental, and the like.
- Analyze only "significant risk events and their relationships," which are the risks that can make a difference in the project cost and schedule. All too often risk analysis loses itself "in the weeds" by focusing on minutiae.

The last bullet deserves a short digression from our regular course since many times the authors have reviewed RBE reports that included many insignificant variables (uncertainties and risks). We would like to bring to readers' attention the message contained in the next two cartoons. Cartoon 3.1 shows the situation when risks are defined as little bits (little rocks and dirt) and risk assessment is a waste of time, since at the end of day, after the dust has settled, a pile was built naturally with or without risk response planning.

Cartoon 3.2 shows a different and scary situation. The same volume of rock and dirt as presented in Cartoon 3.1 is now concentrated on a large rock. In this situation, risk assessment is critical and if the risk response is not implemented in a timely manner, then the only response is running away from it.

The big rock may produce substantial damage to the project while the little rocks may affect the project somehow but will not derail it. Now let's continue with our subject project.

The subject project was in its infancy (less than 5 percent design level) and the scope of the project was not well defined. Project management decided to break the estimate of this large project into multiple contracts and conduct separate cost risk workshops for each contract. This decision was detrimental to the process for the following reasons:

- It led to unnecessary fragmentation of the estimate.
- It created inconsistency with respect to the risk elicitation criteria.
- It developed a sense of false precision to the entire analysis since the data available did not support it.

Risk assessment
and analysis

Risk response

Cartoon 3.1   Risk Management of Many Insignificant Risks

Risk assessment
and analysis

Risk response

Cartoon 3.2   Risk Management of a Significant Risk

In other words, breaking the estimate into contracts may seem like a logical approach but is detrimental to risk analysis as will be illustrated in the next chapters. It is important to "Keep It Simple Smarty" and to analyze the project based on the quality of the data that feed the model. Breaking down the estimates of large projects into many possible contracts (professional sophistication) may be an engaging and useful exercise in theory but it usually lacks the transparency and consistency in practice.

For the previous project, a large number of risks were elicited and analyzed within the model. One danger that was introduced by the presence of a large number of relatively minor risks was the possibility of including risks already captured under base uncertainty. (In this case the base uncertainty is nothing else than assigning a range to the estimated deterministic value. It may represent the base variability that the basis of estimate document will present later. Some professionals use *variability, uncertainty,* and *risks* as interchangeable. The book will clarify and make a distinction between base variability and base uncertainty.) The *uncertainty* in the base alone has created significant range for the base cost. When large uncertainty in the base cost or schedule is used there is a real possibility that the model may double the impact of many nonsignificant risks and create a significant and unjustified shift to the right of the entire cost distribution.

The worst-case scenario is created when a large base uncertainty is justified by different events that may happen but they are not identified. In other words, risks are considered but not identified and quantified. Then when risks are elicited during risk elicitation the risks hidden in base uncertainty may very well be captured and included in analysis. So the analysis will double their effect.

In addition to that, when risk analysis introduces hundreds of risk events and the workshop is overloaded with discussions about all of them, an inordinate amount of workshop time is spent on the discussion of minor issues, which diminishes the time spent on significant risks. Based on observing hundreds of workshops, spending too much time discussing relatively minor issues is a habitual problem and difficult to remedy. People tend to feel more comfortable in discussing little things that have little impact rather than discussing serious events that could derail the project. This tendency is likely related to a phenomenon known as *denial bias*, one of the many types of cognitive biases that will be explored later in this book.[5]

Professional sophistication encourages this kind of behavior by creating a discussion framework. KISS introduces an alternate discussion framework that focuses workshop discussions on issues that really matter.

Dependency among Risks Analyzed

An effective risk analyst knows that accurately capturing the dependencies between risks is crucial for producing reliable and defendable results. Having identified, quantified, and analyzed hundreds of risks dramatically reduces the chance of capturing the right relationship between them. So in many cases the risks are considered independent of each other with dire consequences to the final results. Usually, if risks are considered independent of each other, they will cancel each other out. The following paragraphs present examples of risks elicited and quantified for a $1 billion project where the dependency among risks was ignored.

TABLE 3.4   Base Cost Uncertainty Only vs.
Base Cost Uncertainty Plus Risks

Percentage	Base Cost Uncertainty	Delta Base Cost Uncertainty	Total Base Cost Uncertainty + Escalation + Event Risks	Delta Total
0%	$441		$355	
10%	$660	-27%	$787	-29%
20%	$748		$898	
30%	$808		$975	
40%	$860		$1,047	
50%	$907		$1,108	
60%	$958		$1,182	
70%	$1,014		$1,259	
80%	$1,084		$1,350	
90%	$1,172	29%	$1,464	32%
100%	$1,405		$2,046	

Note: Delta is calculated as a ratio to median.

Table 3.4 presents the base cost uncertainty only and base cost uncertainty and risks altogether using two distinct columns. It indicates the percentage of not exceeding an estimated value. The table may be read such as "there is 10 percent probability of the base cost uncertainty not exceeding $660 million or there is 90 percent probability of the base cost uncertainty exceeding $660 million."

The authors have interpolated two additional columns (Delta Base Cost Uncertainty and Delta Total) that indicate the variance in percentage of rows' values to median value. While the values presented in columns Delta Base Cost Uncertainty and Delta Total indicate that a reasonable range exists, the overall risks' impact consists of increasing the cost of the project by $200 million (22 percent) at the median level, but unexpectedly the risks provide a 2 to 3 percent increase of the "LOW" or "HIGH" delta values relative to the median. The fact that the distribution's margins are not significantly changed by the addition of risks indicates that there is a problem with the analysis.

The general expectation is that risks (threats and/or opportunities—later in the chapter better "risk" definitions are presented) push the ends of the tails of the cost or schedule distribution farther away from the median. For the subject project, despite the fact that the project had significant threats and opportunities, the model didn't predict a noticeable expansion of its

TABLE 3.5   Risks Dependency Relationship

Risk's Name	Risk's Description	Probability of occurrence	LOW [$M]	Most Likely [$M]	HIGH [$M]
Contractor underbids	Qualified contractor underbids the job and defaults on the contract	10%	100	300	500
EPB TBM	Contractor demonstrates ability to use EPB	50%	-40	-35	-30
Opportunity sharing procurement	Procurement strategy to include risk sharing allocation to attract additional bidders and reduce the risk for the owner.	75%	-150	-100	- 10
Design-builder innovation	Design-builder innovation reduces time and cost	80%	-80	-50	-10

tails. One of the reasons is the fact that all risks were treated independently and the effect of opportunities was neutralized by the risks' impact.

For example, Table 3.5, which represents a fraction of the project's risk register, shows a huge threat, *contractor underbids,* and three large opportunities. Besides the fact that they have faulty and/or incorrect descriptions, they are considered independent of each other. These risks should be dependent and their relationship modeled accordingly.

For example, it is hard to accept that a contactor who is saving approximately $185 million (approximately 20 percent of the contract value) is going to default at a rate of 1 out of 10. Perhaps a better risk assessment may reveal that the threat is mutually exclusive with the opportunities. This correction alone may push the ends of the tails farther away from the median value. Risk analysis is about examining the tails. The project manager needs to focus on the upper tail, for therein lies the most significant events that lend themselves to cost overruns and/or delays.

Definition of the Risk Distributions

Risks elicited during the workshop were given impact values by "low, most likely, and high" and translated into the model using a *Trigen* distribution, as shown in Figure 3.8. It looks like the KISS principle was applied. Only in this case the "Keep It Simple, Smarty" was changed to a "Keep It Simpler, Stupid" approach and the results of this simplification may have undermined the original intent of the subject matter experts. Figure 3.8 uses units in millions of dollars.

The workshop data (risk register data) indicate that the low value represents the 20 percent probability of occurring and the high value represents the 80 percent probability of occurring. That means that risk definition allows 40 percent of values used during simulation to lie outside of the range defined by the LOW and HIGH. This approach helps in increasing the final range

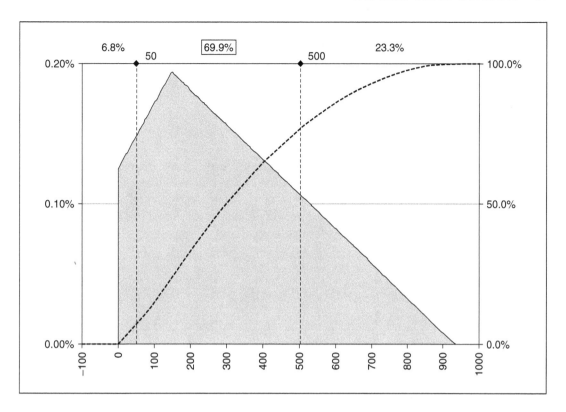

Figure 3.8 Trigen Distribution

but it raises questions about the integrity of translation between workshop data and model data. For example, Figure 3.8 provides information about how the model uses a risk defined in the workshop by its impact: Low = $50 M, Most likely = $150 M, and High = $500 M.

The previous risk is a threat to the project so it cannot have negative values. To accommodate this condition the modeler appropriately truncates the risk at zero, but the truncation produces an undesirable change of the risk's definition. The software changes the meaning of LOW and HIGH from 20 and 80 percent to 6.8 and 76.7 percent, respectively. It means that the function employed by the model allows 30.1 percent of values used during simulation to lie outside of the range defined by the workshop members.

Furthermore, the most disturbing consequence of 20 and 80 percent margins is the fact that the model picks up values between $0 M and $933 M. Figure 3.8 clearly indicates that 23.3 percent of the time the analysis uses values greater than $500 M for this risk. Knowing that the model selects values outside of its definition, someone may ask the question: "Is this what the subject matter experts were trying to articulate?" Probably not. We can almost guarantee that none of the workshop members thought about $900 M.

The accuracy of translation between the workshop data and model inputs is a very important issue. Many times the authors have noticed discrepancies between what the SMEs had in their minds and what the model used. Fortunately, there is a simple solution for avoiding these mishaps. It is recommended that during the workshop, the real range of the risk be displayed in front of the SMEs so everyone understands it.

Professional Sophistication Summary

When professional sophistication runs high the possibility of having issues as previously described is also high. While dramatic failures created by so-called typos are rare, they may happen and every risk analyst should be aware of it.

The most detrimental effects to the RBE process are generated by hidden components of professional sophistication such as: (1) a large number of variables (risks), (2) poor risk conditionality (dependency and/or correlation), and (3) vague definition of a risk's distribution.

While KISS may guard the analysis against some of these fallacies, the KISS principle is not a panacea for risk analysis. The role of the risk analyst and/or risk elicitor is crucial on any risk assessment and risk analysis.

RISK-BASED ESTIMATE—HOW IT WORKS

The previous information was intended to put readers on guard and advise them about the possible pitfalls that RBE may present if the process is misunderstood or misused. The misunderstanding may be cured by the users' desire of learning. One of the book's goals is to provide information to the readers that have been gathered over several years of performing RBEs. The misuse of the RBE process is perhaps the most dangerous issue that may occur and is also the most difficult to correct. It is important for users to understand the concept and components of the RBE process since, at first glance it looks easy, but the reality is that developing proficiency with RBE is a significant undertaking.

Figure 3.9 presents the main algorithm of the RBE process. It is simple but requires a thorough understanding and consistency in its application.

The process starts with examining the existing estimate, sometimes called the *engineer's estimate*. The engineer's estimate (EE) may have been developed for the first time for the benefit of the workshop, or perhaps was previously prepared, and should include all cost elements that affect the project. The EE is typically prepared in the form of a deterministic estimate that follows the appropriate methodologies as described in Chapter 2. The EE usually includes contingencies in some manner, whether explicitly or not.

A team of "outside sagacious estimators" in collaboration with the project estimators will validate the EE though a process called *validation of the base cost and schedule*. This process will be described later. Based on the quality of the data included in the estimate, the base cost and schedule team will recommend the range of variability in the base. Variability in the base and uncertainty in the base will be discussed in greater detail later on.

After the base cost and schedule are validated and the base variability is established, the workshop will focus on discussing the significant risk events, which may change the project cost and/or duration beyond the limits given by the variability in the base. This activity is called *risk elicitation* and it must be facilitated by an experienced risk elicitor.

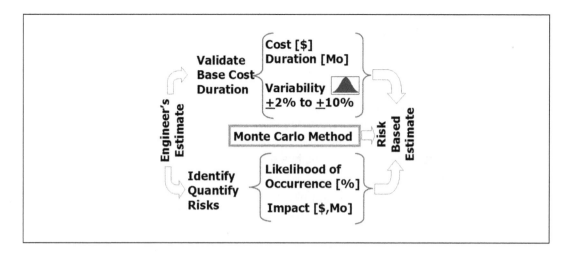

Figure 3.9 Risk-Based Estimate—The Process

After the base cost and schedule are validated and risks are quantified, the Monte Carlo Method (MCM) is used by running thousands of plausible cases and developing a database of possible outcomes. The database created will serve as a resource for developing histograms, tables, cumulative distribution functions, and tornado diagrams that will help communicate the cumulative effect of all identified risks and assist the decision makers to better understand the project and its risk environment.

BASE COST AND SCHEDULE

The review and validation of the base cost and schedule estimate is conducted by the cost lead, who is an experienced estimator with no stake in the project. The objective of cost lead in performing this task is to develop an estimate assuming neutral conditions. It is critical that the values of the estimated base cost and schedule be as accurate as project conditions warrant as discussed in Chapter 2. The estimated base cost and schedule represent the cornerstone of a project's estimate, and any error would induce linear errors in the project's total estimate. This process must be commensurate to the level of knowledge about the project that exists at that time. It is important to follow the methodology presented in Chapter 2.

The review process has the following major steps:
1. Review and validate the basis of estimate (project assumptions).
2. Review project cost and schedule.
3. Remove contingencies hidden or explicitly presented.
4. Capture the "unknown cost" of miscellaneous items.
5. Assign variability to the base estimate.

Detailed descriptions:

1. Review and validate project assumptions.

    Examine, discuss, and document the basis for the estimate. This step is critical because it helps the workshop participants better understand the project (scope and schedule). The outcome of this step will lay the foundation for risk elicitation.

2. Review the project cost and schedule.

    During this step, the workshop team reviews the unit price and quantities and updates them as needed based on the information available. It is very important to document any changes that are made to the base estimate.

3. Remove contingencies.

    The group focuses on removing all contingencies which are hidden or included in various items to cover "what-if" scenarios or risks. At this time, all group members understand that the contingencies removed will be replaced by clearly identified and quantified risks. An item called *change order contingency* remains and it covers minor omissions or errors in the plans.

4. Capture the "unknown cost" of miscellaneous items.

    The unknown cost of miscellaneous items is usually called a *design allowance*. This step covers the cost of items that are included in the project but, at the time of estimate, there is little or no data about how much they might cost.[6]

5. Assign variability to the base estimate.

    The neutral effect of variability should be preserved. The authors recommend using the symmetrical form of either a Pert or triangular distribution, which better captures the meaning of variability. Figure 3.10 presents a base cost of $10 million with a "±10 percent variability." The variability presented shows the absolute limits of the distribution range. It means that the base cost through its variability cannot be lower than $9 million and cannot be higher than $11 million.

    There are situations when the risk elicitor feels more comfortable with assigning relative limits to base variability. For example, assuming that the LOW value represents the 10 percent probability of occurring and the HIGH value represents the 90 percent probability of occurring allows that 20 percent of the values used during simulation are outside of the range defined by the LOW and HIGH. This approach has undesirable side effects on analysis results as it was presented in previous examples.

In summary, the validation of the estimated base cost and base schedule ensure that: (1) the assumptions and basis of the estimate are appropriate for the project; (2) items are not missing; (3) historical data, the cost-based estimate, or other data that were used to develop the estimate accurately reflect the project scope and site conditions; and (4) the base cost and schedule estimates are accurate reflections of the project's scope of work.

Base Uncertainty (Probability Bounds, p-box)

A project's base cost and schedule estimate basically comprises two major types of uncertainties: (1) epistemic (lack of knowledge) uncertainty, and (2) random uncertainty.[7]

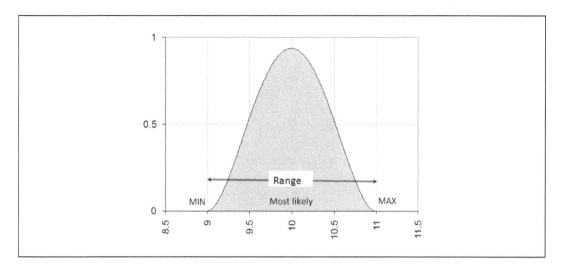

Figure 3.10   Base Variability "$10 M ± 10 percent"

Epistemic uncertainty is produced by the lack of knowledge about the project and its magnitude may always be reduced by acquiring information on the project and/or by consulting with the subject matter experts. Epistemic uncertainty, also known as incertitude, ignorance, subjective uncertainty, nonspecificity, or reducible uncertainty, may be easily reduced by acquiring knowledge on the subject.[8]

The epistemic uncertainty component of the base cost and schedule estimate is named in this book as *variability* in base cost and schedule estimate, and it was presented in preceding paragraphs. Figure 2.2 captures the reducible nature of the variability component for the estimated base cost and schedule. Variability in the base estimate shrinks as the design evolves and more data about the project becomes available. During the design phase after the project acquired a definite definition, the project's base cost uncertainty depends on the quality of data related to things such as the cost of materials, equipment, and labor.

It is important to notice that the variability as epistemic uncertainty has a symmetrical distribution. The lack of knowledge may go either way: increase the estimated value or reduce the estimated value. Any asymmetry needs to be captured by risk events. As a reminder, base cost and schedule are defined for eventless conditions.

The second component of the uncertainty in the base estimate has a random nature and is produced by uncontrollable changes at the time of advertisement. Usually this type of uncertainty is known as random uncertainty, stochastic uncertainty, objective uncertainty, dissonance, or irreducible uncertainty as it may not be reduced since it is generated by elements such as: acts of nature, higher level decision, market conditions, and so forth.

The Epistemic Component — Base Variability

There are divided opinions about how variability in the base should be assigned to the base cost and schedule. The authors examined hundreds of RBE reports provided by different professionals and a wide array of base variability implementation was observed.

There are professionals who assign range and shape to any item that contributes to the base component; as it was demonstrated before this is a bad practice because it leads to an unreasonably narrow range if they are not positively correlated. In many instances the range assigned to the base or base component is asymmetrical (−10 percent, +20 percent). It appears that professional sophistication is at its highest because the variability distribution has its own shape produced by sophisticated thinking. Let us first clarify why asymmetrical distribution of base variability is not just unnecessary complication but it is detrimental to the process.

Base Variability — Symmetrical versus Asymmetrical Distribution

The authors' experience with hundreds of risk workshops and the findings of their research suggest that the range assigned to the base variability should be symmetrical with no exceptions. Some professionals are saying that the asymmetry of base variability is necessary because asymmetry expresses what the SMEs think. We have heard professionals saying, "Designers and contractors are telling me that the cost of a specific activity may range from $10 million to $30 million with most likely $12 million, so I have to listen to them and place in the model whatever they think may happen." This is seemingly a reasonable statement; but is it?

The fallacy of this logic stems from the fact that risks are included on the SME's estimated activity cost. RBE is a relatively new way of estimating and most SMEs' experience is related to the global cost (base plus risks) and it is the responsibility of the base cost lead and risk elicitor to ensure that the base cost and schedule *do not include any risks.*

The previous example may have a scenario like this: base cost $12 M with variability of ±20 percent and a threat with 50 percent probability of occurrence and impact of: minimum = $0.4 M, most likely = $4 M, and maximum = $15.6 M. Figure 3.11 displays the cumulative distribution functions of asymmetrical large base uncertainty versus base variability complemented by risks as defined in this paragraph.

Figure 3.11 shows that the two approaches lead to similar results, but having risks identified and quantified creates a means for risk management. RBE is an estimating process that produces usable data for project risk management. Chapter 6 will present a second example of how large asymmetrical base uncertainty may be replaced by base variability and risks.

Next we will present the detrimental effect of asymmetric variability for the most used distribution: (1) Pert, (2) triangular, and (3) uniform distributions.

Pert Distribution

Figure 3.12 shows what kinds of changes are created when an asymmetrical distribution replaces a symmetrical one. This figure represents how the cost is distributed when the distribution range is symmetrical and asymmetrical. Both distributions have the same "most likely" value (100). The symmetrical distribution is defined by ±10 percent for the minimum and maximum range values and the asymmetrical distribution is defined by −10 percent for the minimum value, and +30 percent for the maximum value.

The cumulative distribution function shows a clear increase of the base cost at the higher confidence level. At a high level of confidence this increase may reach 10 percent or higher. For

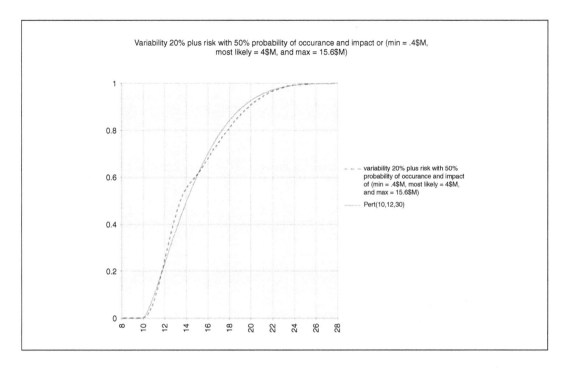

Variability 20% plus risk with 50% probability of occurance and impact or (min = .4$M,
most likely = 4$M, and max = 15.6$M)

– – – variability 20% plus risk with 50% probability of occurance and impact of (min = .4$M, most likely = 4$M, and max = 15.6$M)

——— Pert(10,12,30)

Figure 3.11   Simple Large Asymmetrical Base Uncertainty versus Combined Symmetrical Variability and Risks

example, at 70 percent confidence level, which many organizations often choose for establishing budget numbers, the asymmetrical distribution produces an increase of 5 percent on top of the increase provided by symmetrical distribution. Overall, at 70 percent confidence level, the asymmetrical distribution provides 7.5 percent increase in the base cost values related to its deterministic value.

Furthermore, at a 90 percent confidence level the cost added by asymmetrical distribution is increased by 10 percent on top of the 5 percent increase given by the symmetrical distribution. So at 90 percent confidence level the base cost will have a value of about 11.5 percent of its deterministic value.

Moreover, the mean and median values increase by an average of 3 percent. The increase of the mean and median values suggests a hidden shift of the base cost toward the higher numbers and this shift generates the most troublesome effect that relates to having undocumented and unjustified change of the base values. Perhaps the majority of users (project team members, project managers, subject matter experts, base cost lead, and sometimes the risk leads) are not aware of the real implication that the asymmetry in the base produces. Further discussion about the implication of using asymmetrical distribution to reflect variability in base values follows.

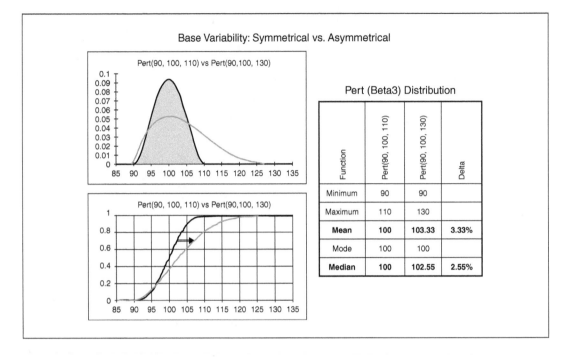

Figure 3.12   Pert Distribution: Symmetrical versus Asymmetrical

An important concept to remember throughout the entire risk-based estimating process is that the user must be able to explain the numbers and the effects of risks and variability should be narrative.

Triangular Distribution

This trend of change in base values is more significant if the distribution has a triangular or uniform shape, as shown in Figures 3.13 and 3.14. Figure 3.14 clearly shows that the asymmetry of the triangular distribution has a significant impact on overall distribution values. The cumulative distribution functions (CDFs) of the triangular distribution do not intersect while the CDFs of the Pert distributions intersect. On the other hand, the asymmetry on triangular distribution decreases the density values at the lower end and increases the density values at the higher end. The overall effect is a more dramatic shift of the base values toward the high end of its range.

In the case of triangular distribution, the increase of the mean and median values doubles if it is compared with similar values given by a Pert distribution. At the same time, the increase of base cost value at a higher confidence level is larger. For example, at a 70 percent confidence level, the asymmetrical distribution adds 9 percent on top of the increase provided by a symmetrical distribution. Overall, at 70 percent confidence level the asymmetrical distribution provides 12 percent increase in the base cost values.

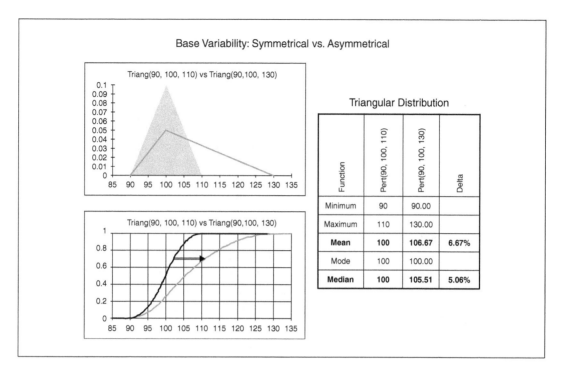

Figure 3.13   Triangular Distribution: Symmetrical versus Asymmetrical

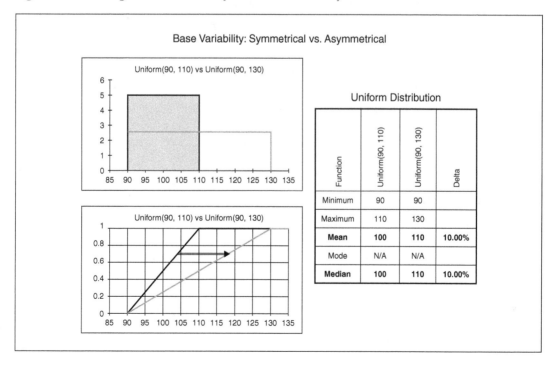

Figure 3.14   Uniform Distribution: Symmetrical versus Asymmetrical

Uniform Distribution

The most dramatic example is provided by using a uniform distribution to represent the uncertainty in the base. It is true that the uniform distribution has just two parameters (LOW and HIGH) but the issue of asymmetry is still valid. The base value could be 100 units ($ or days) and, using the concept of variability in base, someone may consider that the uncertainty is in the form of symmetrical uniform distribution (90, 110) or asymmetrical uniform distribution (90, 130). So in the case of uniform distribution, the asymmetry is related to the base deterministic value.

In the case of uniform distribution the change is dramatic. The increased values of the mean and median are as high as 10 percent. The increase is significantly higher for high confidence levels. The uniform distribution presents an unambiguous case of the damaging and misleading effect that an asymmetrical distribution of base variability produces on risk analysis results.

Summary of Finding Related to Base Variability

The most important finding of the analysis presented above is the fact that the use of an asymmetrical distribution to express variability in the base cost or duration will change the meaning of the definition of base cost or schedule estimate. Simultaneously, any change in the base cost and/or duration values accomplished by employing an asymmetrical distribution is done under a "hidden condition." The authors assume that SMEs are likely to be unaware of the hidden effect of employing an asymmetrical distribution to represent the uncertainty in the base cost or schedule.

The majority of SMEs have developed a bias toward seeing costs represented by an asymmetrical distribution, but they have forgotten that their experience is related to project costs in general, and that includes risks in the form of a standard, flat contingency factor. The base cost, by its definition, must be free of any significant risk-related events that would skew its distribution.

As a reminder: The base cost is the estimated cost value when the project is delivered as planned. The variability in the base is designed to capture the uncertainty in the cost of material, equipment, and labor in the conditions when no events will disturb the project delivery.

While the variability in base is an important component of the RBE, capturing the market conditions in cost and schedule risk analysis is essential for defining a reasonable and healthy range of the estimate.

Base's Random Components — Market Conditions

Market conditions are of special interest in the analysis of cost and schedule estimates since they may produce a significant impact on their outcome. Market conditions may be viewed as the "random uncertainty" component of the base estimate's uncertainty. Market conditions are generally uncontrollable; however, project costs might be managed considering market conditions in order to optimize them.

While market conditions may not be controllable by ordinary means, sometimes a good project manager may reduce their impact. When the project schedule allows for a wider window of the project's start date, market conditions may be addressed to some degree by controlling the construction letting date. There are two main scenarios: (1) avoiding high prices by delaying or accelerating the start of the construction, or (2) sequencing the project construction with favorable market conditions.

Considering that the base estimate is an aggregate of base variability and market conditions, the base estimate uncertainty may be represented by the probability bounds approach which blends together epistemic and random uncertainty by creating a probability box (p-box). Probability bounds have been used in different forms by Ferson and Hajago,[9] Tucker and Ferson,[10] Bruns and Paredis,[11] and Frey and Bharvirkar[12] by applying it to environmental risk assessment, and in the analysis of imprecise uncertainty.

The terms such as *variability* and *uncertainty* have different meanings for different professionals. The book presents these terms in relation to base cost and schedule. Chapter 2 defines base variability and this chapter presents definitions of base uncertainty. The key word of our definition is "base."

It is notable that the concept of probability box is presented in a little different form by Vose.[13] His book *Risk Analysis—A Quantitative Guide* presents: (1) variability as a random variable, (2) uncertainty as an epistemic variable, and (3) total uncertainty as a combination of the first two. The concept presented in this book is similar but with a twist on the nature of terms. We consider base variability as epistemic in nature, market conditions (letting conditions) as a random variable, and base uncertainty as a combination of the first two.

The following paragraphs will present how the probability box can be employed for treating base cost and schedule variability and market conditions as a structure that allows the comprehensive propagation of both random and epistemic uncertainty through the MCM. Figure 3.15 presents a schematic of how the cost base uncertainty p-box may be defined.

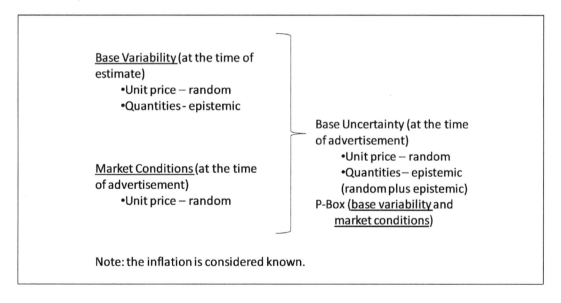

Figure 3.15   Base Uncertainty p-Box

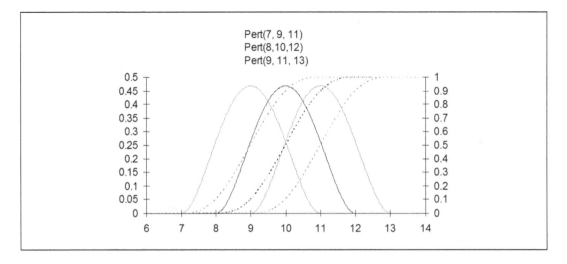

Figure 3.16   Uncalibrated Base Uncertainty p-Box

Consider a project that is estimated to $10 million and for this exercise we have assumed the project base cost is represented by a Pert distribution with a minimum value of $8 M; a maximum value of $12 M; and a most likely value of $10 M. This distribution represents the epistemic part of uncertainty in the base cost and can occur at "any time" during the simulations. It is the base variability and will always have symmetrical shape, as discussed in the previous paragraphs.

The second component of the uncertainty in the base represents the random uncertainty created by market conditions, which may change the most likely value of the base estimate. Assuming that market conditions may lower or increase the most likely value by 10 percent, the p-box components of the base cost uncertainty are represented by Figure 3.16.

Figure 3.16 shows the base uncertainty components in a nonrealistic mode since, by its nature, the base cost could not have more than one value at a time; however, it shows that the shapes of cost distribution are identical. Better-or worse-than-planned market conditions use the same shape distribution. The absolute range value of each distribution is identical ($4 M); the only change is in the most likely value. The most likely value of better-than-planned market conditions is $1 M less than planned conditions. The most likely value of worse-than-planned market conditions is $1 M more than the as planned condition.

In order to bring this p-box to reality—a true representation of base cost uncertainty—it is necessary to calibrate it by applying the probability of occurrence. It is necessary to assign to the better-than-planned market conditions a percentage that will represent the probability of this condition happening and another percentage (which may be equal to the first one) for the worse-than-planned market conditions. Of course, the "as-planned" market conditions will have their own probability of occurrence:

{100% – (better-than-planned percentage) – (worse-than-planned percentage)}

In the next few paragraphs we will examine three probability boxes:

1. A probability box defined by:

$$p\text{-box} = \{10\% \text{ Pert } (7, 9, 11), 70\% \text{ Pert } (8, 10, 12), 20\% \text{ Pert } (9, 11, 13)\}$$

that represents:

- A 10 percent chance that the entire distribution of the base variability slides downward 10 percent from the most likely value of the as-planned distribution—better than planned

- A 20 percent chance that the entire distribution of the base variability slides upward 10 percent from the most likely value of the as-planned distribution—worse than planned

- A 70 percent chance that the base variability distribution will not move at all

2. A probability box defined by:

$$p\text{-box} = \{10\% \text{ Pert } (8, 8.1, 12), 70\% \text{ Pert } (8, 10, 12), 20\% \text{ Pert } (8, 11.9, 12)\}$$

that represents:

- A 10 percent chance that the most likely value of the better-than-planned market condition distribution moves very close to the planned distribution's low end but its range stays unaffected—better than planned

- A 20 percent chance that the most likely value of the worse-than-planned market conditions distribution moves very close to the as-planned distribution's high end but the range stays unaffected—worse than planned

- A 70 percent chance that the distribution will not change

3. A probability box defined by:

$$p\text{-box} = \{10\% \text{ Pert } (7, 7.1, 11), 70\% \text{ Pert } (8, 10, 12), 20\% \text{ Pert } (9, 12.9, 13)\}$$

that represents:

- A 10 percent chance that the better-than-planned market condition distribution's range slides 10 percent downward from the planned distribution and the most likely value of the better-than-expected market condition moves in the proximity of its new low end—better than planned

- A 20 percent chance that the worse-than-planned market condition distribution's range slides 10 percent upward from the planned distribution and the most likely value of the worse-than-expected market condition moves in the proximity of its new high end—worse than planned

- A 70 percent chance that the distribution will not change

Figure 3.17 represents the p-box components of the base cost uncertainty defined by the

$$p\text{-box} = \{10\% \text{ Pert } (7, 9, 11), 70\% \text{ Pert } (8, 10, 12), 20\% \text{ Pert } (9, 11, 13)\}.$$

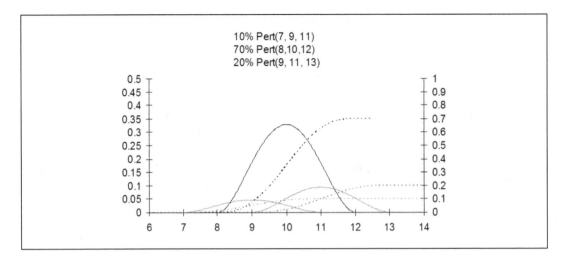

Figure 3.17   p-box = {10% Pert (7, 9, 11), 70% Pert (8, 10, 12), 20% Pert (9, 11, 13)} Base
Uncertainty—Entry Distributions

Figure 3.17 looks similar to Figure 3.16 with the difference being in the magnitude of most
likely frequency values of each distribution.

The vertex of each distribution shown on the left axis is at half of the probability of the
occurrence for the respective distribution. The maximum value of each cumulative distribution
function indicates the probability of occurrence of that distribution as is shown on the right axis.
The right axes may be useful in maintaining quality control of how the p-box is defined because
the sum of the maximum values of each function will be equal to one.

Figure 3.18 shows the results provided by running the previous p-box through 10,000
iterations. One notable observation is that the lower and higher end margins extend significantly
for the better- and worse-than-planned market conditions. The fact that the worse-than-planned
market conditions have a higher probability of occurrence moves the graph toward the right and
breaks the symmetry of the base cost uncertainty.

Figure 3.19 represents the p-box components of the base cost uncertainty defined by the
p-box = {10% Pert (8, 8.1, 12), 70% Pert (8, 10, 12), 20% Pert (8, 11.9, 12)}. It represents an
unusual situation whereby market conditions push the estimated cost toward the distribution's
ends but do not extend the distribution's boundaries. This kind of situation requires advanced
modeling and may be useful when market conditions suggest that the values of the distribution
range of the base variability is okay but thicker tails are needed.

Figure 3.20 shows the results provided by running the p-box = {10% Pert (8, 8.1, 12), 70%
Pert (8, 10, 12), 20% Pert (8, 11.9, 12)} through 10,000 iterations. The effects of the better-
than-planned market conditions and the effect of the worse-than-planned market conditions are
worth noting. The distribution tails do not extend since this was an input condition, however, the
probability increases substantially at both tails. The fact that the worse market conditions have a
higher probability of occurrence moves the graph toward the right and breaks the symmetry of
the base cost uncertainty.

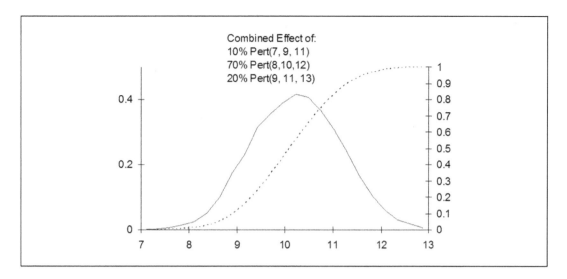

Figure 3.18 p-box = {10% Pert (7, 9, 11), 70% Pert (8, 10, 12), 20% Pert (9, 11, 13)} Base Uncertainty—Simulated Effect

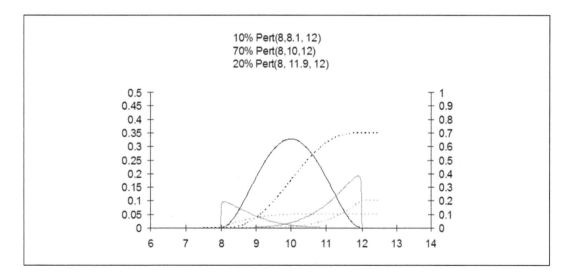

Figure 3.19 p-box = {10% Pert (8, 8.1, 12), 70% Pert (8, 10, 12), 20% Pert (8, 11.9, 12)} Base Uncertainty—Entry Distributions

Figures 3.19 and 3.20 show an extreme situation when the most likely values of the better-than-planned and worse-than-planned conditions are very close to the planned base variability's ends. Of course, the most likely values for the market conditions could be anywhere inside the distribution range and the histogram could have a smoother shape. It is a matter of how the SMEs consider market conditions.

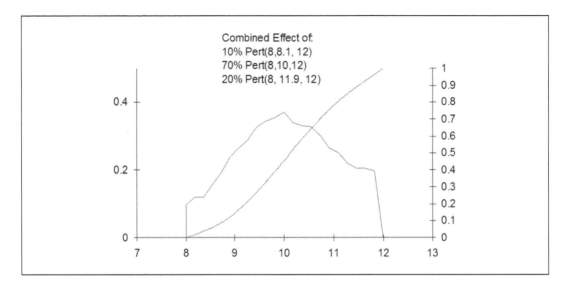

Figure 3.20 p-box = {10% Pert (8, 8.1, 12), 70% Pert (8, 10, 12), 20% Pert (8, 11.9, 12)} Base Uncertainty—Simulated Effect

Figure 3.21 represents the p-box components of the base cost uncertainty defined by the p-box = {10% Pert (7, 7.1, 11), 70% Pert (8, 10, 12), 20% Pert (9, 12.9, 13)}. This represents a situation when market conditions push the estimated cost toward the distribution's tails and at the same time extends the distribution's boundaries. In other words, Figure 3.21 represents a combination of the first two cases. It may be used for cases when significant market conditions are forecast.

Figure 3.22 shows the results provided by running the p-box = {10% Pert (7, 7.1, 11), 70% Pert (8, 10, 12), 20% Pert (9, 12.9, 13)} through 10,000 iterations. There are significant changes in the p-box distribution where, as forecasted, the lower end and higher end of the distribution are quite relevant. There is a dominant effect of the better-than-planned market condition and of the worse-than-planned market condition. The distribution tails extend and the probability increases substantially at both tails. The fact that the worse than planned market conditions have a higher probability of occurrence moves the graph toward the right and breaks the symmetry of the base cost uncertainty.

Base Uncertainty and Noise

As has been discussed so far, base uncertainty has a robust and sound definition. The primary concern is how it can be measured and how to avoid the related pitfalls. It is recommended that the risk elicitor, base cost lead, and some key SMEs establish the magnitude and probability of market conditions. The only recommendation relevant to base uncertainty is that any RBE analysis should include the following scenarios: (1) better than planned and (2) worse than planned. Nobody has a crystal ball.

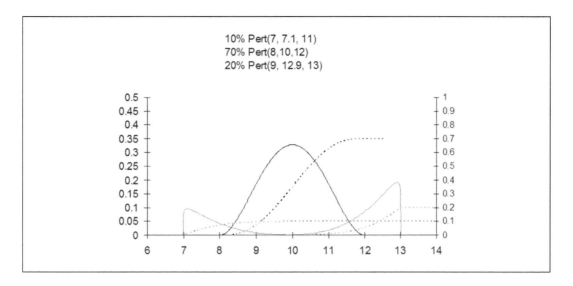

Figure 3.21   p-box = {10% Pert (7, 7.1, 11), 70% Pert (8, 10, 12), 20% Pert (9, 12.9, 13)} Base Uncertainty—Entry Distributions

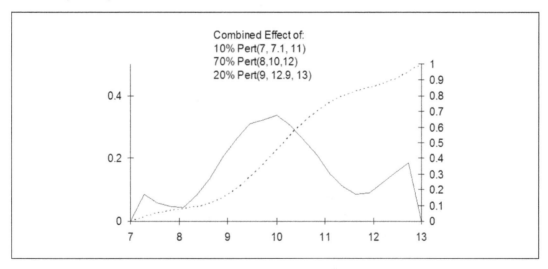

Figure 3.22   p-box = {10% Pert (7, 7.1, 11), 70% Pert (8, 10, 12), 20% Pert (9, 12.9, 13)} Base Uncertainty—Simulated Effect

Care should be taken when the magnitude of base variability is chosen. The authors have witnessed the entire spectrum of base variability: ranging from the deterministic (just one number) to very large ranges (±40 percent). How the base variability magnitude may affect the quality of results provided by RBE analysis is an issue of concern that will be addressed in the following paragraphs.

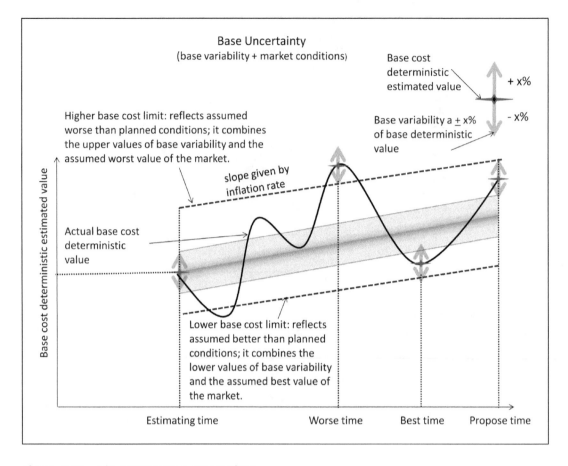

Figure 3.23 Time versus Base Uncertainty

Base uncertainty is represented in Figure 3.23 and it shows the boundary of base variability and market conditions forecast over time. The solid continuous line represents just one of the possible scenarios that may happen. As can be seen here, regular inflation has nothing to do with base uncertainty. Normal inflation will push the cost higher according to the planned conditions. The graph shows that the magnitude of the "better than planned" scenario is less than the magnitude of the "worse than planned." At the same time, Figure 3.23 shows that the uncertainty in the base may not cover all possible market conditions (where the continuous line intersects the uncertainty's boundaries). The issue of concern here is how large base variability ($\pm X$ percent) should be. Since the base estimate is a relatively new concept, there is no data to relate to. As a reminder to the reader: Base estimate assumes no events during project delivery and as far as we know there is no such of projects.

Recently, the authors have noticed an alarming trend in cost risk analysis whereby the magnitude of base variability values has been increased to the point where they are defeating the purpose the analysis. This trend of increasing the magnitude of the base variability

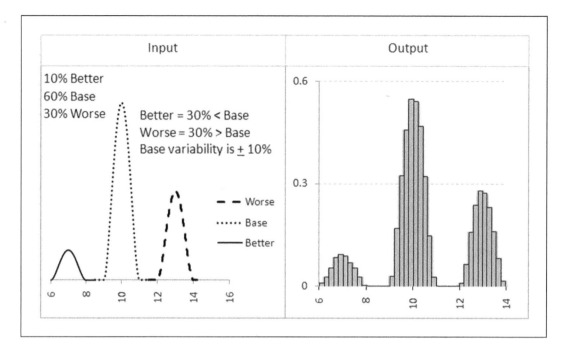

Figure 3.24   Base Uncertainty (Variability ±10 Percent; Market Shift +30 Percent, 10 Percent Probability Better Than Planned; 30 Percent Probability Worse Than Planned)

introduces an umbrella effect that conceals the impact of market conditions and at times risks. It introduces what amounts to a "static" or "noise effect" in the analysis.

Figures 3.24 through 3.27 present the result of base uncertainty for a project that cost $10 million and consider a series of different situations: The first three figures have: 10 percent probability of occurrence of better than planned when the cost will be 30 percent less than planned, and 30 percent probability of occurrence of worse than planned when the cost will be 30 percent higher than planned. The distinction between figures consists in different variability values: Figure 3.24 has ±10 percent variability, Figure 3.25 has ±20 percent variability, and Figure 3.26 has ±30 percent variability.

Figure 3.27 has a 10 percent probability of occurrence of better than planned when the cost will be 20 percent less than planned, and 30 percent probability of occurrence of worse than planned when the cost will be 20 percent higher than planned and a base variability of 20 percent.

The graph shown on Figure 3.24 sends a pretty strong message to the viewer. It's making the case on what is happening when the market realizes better or worse than the planned market conditions. The fact that the graph shows three distinct humps with no connection among them is an indicator that the magnitude of the base variability is too low compared with the magnitude of market conditions.

Figure 3.25 presents the same scenario as in Figure 3.24 except with a change in base variability. In this case, base variability was changed from ±10 percent to ±20 percent. The

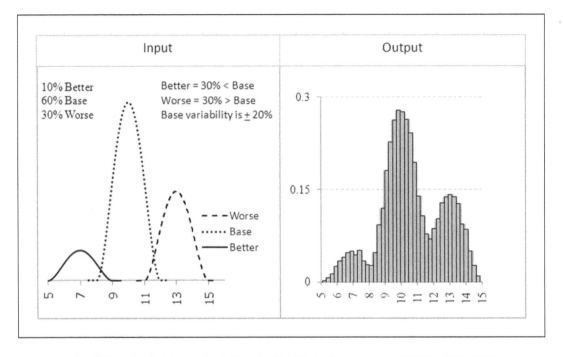

Figure 3.25   Base Uncertainty (Variability ±20 Percent; Market Shift +30 Percent, 10 Percent Probability Better Than Planned; 30 Percent Probability Worse Than Planned)

graph looks better. It is the way it should look—a three-hump histogram, since it represents the base cost of a project when market conditions are significant. Each hump has its meaning and the fact that there is no discontinuity makes the results more practical.

Increasing the base variability from ±20 percent to ±30 percent, the results of the project base cost loses its resolution and only an experienced viewer may see the effect of market conditions (Figure 3.26). It is hard to understand what is going on in the tails and only after reviewing the risk structure someone may associate the tails' shape with market conditions. The histogram resolution becomes worse when the shift in market conditions is equal to or less than the base variability.

The Figure 3.27 represents the situation when the market conditions' probability of occurrence are kept the same as the first three cases presented before but the shift of market is about 20 percent lower or higher than the normal conditions and the variability is ±30 percent. In this case the histogram loses its resolution and the viewer has no idea of what is going on in the analysis.

It is like watching a movie at home in three different formats: Watching a movie in Blue-ray is more pleasing to the viewer's eye and provides the greatest clarity and fidelity; watching it on DVD, while still offering a good picture, the colors are not as crisp and the definition not as sharp; watching it on VCR is a far cry from the first two formats and leaves the viewer underwhelmed.

The same phenomenon essentially occurs when statistical data is displayed. If the shape of data presented doesn't say much because of the lack of resolution, why bother? Of course, it

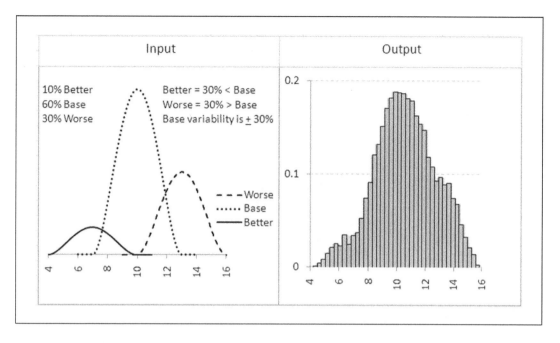

Figure 3.26   Base Uncertainty (Variability ±30 Percent; Market Shift +30 Percent, 10 Percent Probability Better Than Planned; 30 Percent Probability Worse Than Planned

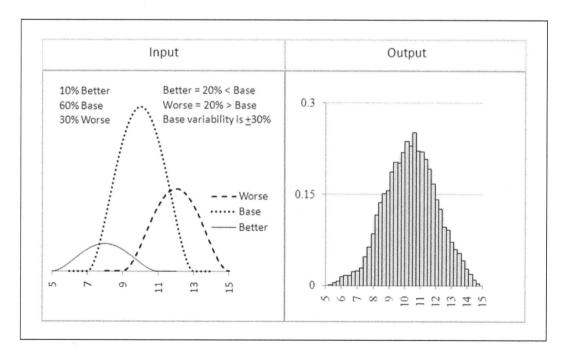

Figure 3.27   Base Uncertainty (Variability ±30 Percent; Market Shift ±20 Percent, 10 Percent Probability Better Than Planned; 30 Percent Probability Worse Than Planned)

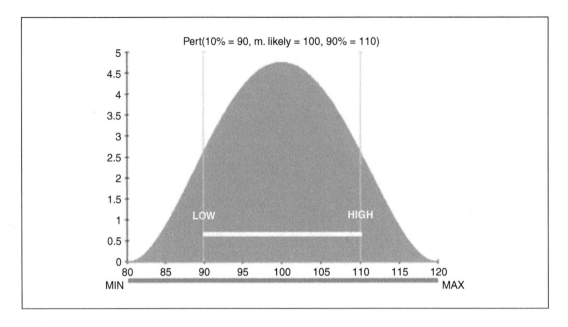

Figure 3.28   Base Variability

must be acknowledged that there may be situations when little can be done to present a vocal shape. A "vocal" shape is the distribution that can tell a story without words.

The issue of *noise* becomes more critical when the variability is defined by terms like "LOW" and "HIGH," where these terms represent symmetrical percentage values such as 10 and 90 percent. Figure 3.28 presents this scenario when there is a 20 percent probability that the numbers used to represent the base cost will be between 80 to 90 and 110 to 120. This means that there is one out of five chances that the numbers selected by the model when it is running its "plausible case" will be outside of the range.

In other words, Figure 3.28 shows that the LOW boundary allows 10 percent of the numbers used during the calculation of the base cost to be below its value and the HIGH boundary allows 10 percent of the numbers used during the simulation to be above its value. This approach, if it is not completely understood, may lead to an analysis that will end with results presented in a form close to a normal distribution. The noise will dominate the analysis and the events that the analysis was supposed to focus on will become invisible in terms of the shape of the data. This makes it difficult for a project manager to communicate the risks impact to others.

Conclusions on Base Uncertainty

The fact must be recognized that at this point in time nobody can know how much a project is actually going to cost five years from now, even if the project is delivered without the occurrence of project-specific events (i.e., risks). We may forecast a range of probable cost (opinion of cost) that is based on the information available at the time of the estimate. This is what base uncertainty is all about.

The probability bound is an excellent concept that captures both components of the base estimate: (1) epistemic and (2) random. Each component is crucial in defining reliable and robust base uncertainty. It is up to knowledgeable professionals to define and calibrate these two components for each individual project. The attention, or lack thereof, given to this step may decide if the analysis will help guide project decision making or hinder it in terms of meeting the project's goals.

RISKS AS EVENTS

There are many definitions for risk, some better than others. An entire chapter could be written about the definition of risk, and still not everyone would agree or be satisfied with the results. The definition of risk was under intense scrutiny before the end of the millennium. In short, the standard accepted definition of risk sounds like: "an uncertainty that, if occurs, will affect the objectives." Pretty short, isn't it? It looks like it wasn't short enough.

The new ISO31000 "Risk Management—Principles and Guidelines" standard (published in November 2009) gives a new, shorter definition of risk: "effect of uncertainty on objectives." Just five words but pretty powerful. The ISO31000 explains each word by adding clarification notes such as:

NOTE 1 An effect is a deviation from the expected—positive and/or negative.

NOTE 2 Objectives can have different aspects (such as financial, health and safety, and environmental goals) and can apply at different levels (such as strategic, organization-wide, project, product and process).

NOTE 3 Risk is often characterized by reference to potential events and consequences, or a combination of these.

NOTE 4 Risk is often expressed in terms of a combination of the consequences of an event (including circumstances) and the associated likelihood of occurrence.

Note 5 Uncertainty is the state, even partial, of deficiency of information related to, under-standing or knowledge of an event, its consequence, or likelihood.

The next two paragraphs present two similar definitions from two different sources.

Exposure to the consequences of uncertainty. In a project context, it is the chance of something happening that will have an impact on objectives. It includes the possibility of loss or gain, or variation from a desired or planned outcome, as a consequence of uncertainty associated with following a particular course of action. Risk thus has two elements: the likelihood or probability of something happening; and the consequences or impacts if it does.

Source: "Project Risk Management Guidelines," 2005 by Cooper, Grey, Raymond, Walker

Project risk—the exposure of stakeholders to the consequences of variations in outcome. The overall risk affecting the whole project defined by components associated with risk events, other sources of uncertainty and associated dependencies, to be managed at the strategic level.

Source: PRAM Guide, 2004 by APM Publishing

For the purpose of this book, a simple, clear definition will be used: *risk is any event that may change the project cost or schedule*. The event by its own definition may or may not occur. So the first measurement of a risk is given by the probability of its occurrence. Also, an event may have different magnitudes. So the second measurement of a risk is its effect on the project. In vernacular terms a risk's effect is called *impact*.

If the event's impact is beneficial to the project (reduces the cost and/or shortens the delivery schedule) the event is called an *opportunity*. When the event is detrimental to the project (increases the cost and/or delays the delivery of the project) the event is called a *threat*. So a risk may be a threat or opportunity and in some cases may be a threat for the cost and an opportunity for the schedule or vice versa.

There is some confusion inside the risk community when people talk about risks and opportunities and associate risk only with threat. The authors have no intention to enforce one way or another regarding the terminology, but for consistency of this book we will stick with the terminology presented earlier.

Risk is an event that must be fully described and, once this is done, anyone should be able to understand what it represents. Too many times risks are poorly described, risks that the same people involved in identifying them couldn't remember several days later. It is recommended that sufficient time be spent on understanding, quantifying, and documenting every single risk that is worthy of analysis.

Nevertheless, it is important that a risk be described following "SMART" principles. The SMART principles are presented in Table 3.6.

Risk's Description

Going back to the previous definition of risk, it may be concluded that a risk is an event that is measured by two characteristics: (1) the probability of its occurrence, and (2) the impact of its occurrence. Defining a risk's probability of occurrence is the most tenuous part of the process of RBE and it is heavily dependent on the expertise of the SMEs.

Risk's Probability of Occurrence

To facilitate the process of assigning a probability of occurrence to a risk event, the authors recommend using the following scale:

- Very Low = 5 percent
- Low = 25 percent
- Medium = 50 percent

- High = 75 percent
- Very High = 95 percent

This scale is for guidance only. In the process of elicitation, any value such as 20 percent (one-fifth), 33 percent (one-third), 67 percent (two-thirds), is acceptable. The value of 0 percent is not acceptable because, in this case, the risk will never occur. It may be called a "non-occurring risk." The value of 100 percent should be avoided as well, but it may be used when a certain element has an unusually wide distribution range that cannot be captured by the base uncertainty.

It is worthwhile to specify the situation when the SMEs have no idea about a risk's probability of occurrence. In this case, the 50 percent value, reserved for *no clue events,* is appropriate. Any assumption used to determine probability of occurrence should be documented.

TABLE 3.6   Risk Elicitation—SMART Principles

SMART

Specific; the event should be specific to the project.

Measureable; the risk description should allow for measurement of the risk and help in measuring its characteristics (probability of occurrence and the impact). It is recommended that the assumptions made during risk identification be captured. Why is the probability given its specific value? What assumptions were made when the LOW, HIGH, or MOST LIKELY value was assigned? Backup calculations are highly recommended.

Attributable; the risk should have an origin. Why may a risk occur? We call this attribute the *risk trigger.* The risk trigger is essential information for the next steps in risk management, monitoring and control.

Relevant; the risk should make a difference in the project cost or schedule. Avoid spending time with minor risks. It is recommended to screen all risks and select for quantitative analysis only those risks that could significantly change the project cost and/or schedule. The relevant risks are the ones worth managing.

Timebound; the risk should have a limited lifespan and should be described in such a way that it allows the project manager to decide when and if a risk should be retired. At the same time, the risk description should provide information alerting the project manger when to intensify his/her watch over the risk.

A good risk elicitation may make some SMEs nervous, but it can bring essential information to the table that improves the quality of the entire effort.

On one hand, at no time should the probability of occurrence be "elicited" to the nearest 1 percent. For example, if the 17 percent probability of occurrence is elicited during workshop, questions will be raised about its precision. This precision is unbelievable and may hurt the credibility of the process.

On the other hand, the probability of occurrence of 17 percent may be associated with a risk if this value is calculated based on interdependencies among events. For example, risk A may depend on the occurrence of risk B in such way as if risk A occurs then risk B may occur with a probability of occurrence of one out of three. If risk A has probability of occurrence of 50 percent, then the probability of occurrence of risk B is 17 percent.

Related to false precision the authors would like to present a quote from a person who proved to the entire world what it means to be reasonable:

> I would rather be approximately right than precisely wrong.
>
> *Warren Buffett*

A short digression from the subject we are having: Quoting Warren Buffett represents a risk for us since we haven't asked permission for it. We use the general acceptance of "limited use." We think that we are dealing with a minor risk. But we never know what may happen.

It should also be mentioned that the first person known to have coined this phrase was actually a British logician and philosopher named Carveth Read who said: "It is better to be vaguely right than exactly wrong."[14]

In summary, it is important that the process of defining the probability of occurrence be as accurate as conditions allow and care should be taken about how much time is spent debating over it. To paraphrase Voltaire: "Waiting for perfection is the greatest enemy of the current good." The RBE focuses on developing the current good, so trust your experts and challenge them on substance and biases; don't negotiate guesses.

Risk Impact

Risk impact is defined by a distribution that is described by its range and shape. Typically, there are two main categories of distributions that may describe a risk's impact:

- Discrete distributions such as:
  - Binomial—The discrete probability distribution of the number of successes in a sequence of n independent "yes/no" experiments, each of which yields success with probability p.
  - Discrete—Each outcome has a value and a probability of occurrence.

Discrete distributions have a limited use in RBE. Table 3.7 presents annotated examples of the most common discrete distributions.

- Continuous distributions such as:
  - Pert distribution—A useful distribution in modeling the RBE process.

TABLE 3.7 Discrete Distributions

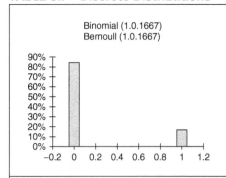

When binomial distribution has just one experiment (yes/no) the binomial distribution is a Bemoull distribution. Rolling a die, where a two is "success" and anything else is a "failure". The die is rolled just once. The value 1 is taken with "success" and value is taken with "failure."

Examples
Did the project include the "A" bridge?
Was the base cost estimate ready?
Did the coin land heads?

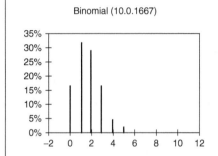

Binomial is the discrete probability distribution of number of successes in a sequence of "n" independently "yes/no" experiments each of which yields success with probability "p".

The histogram represents the situation when the die is rolled 10 times and where a two is "success" and anything else is a "failure."
It is about 16% chance that NO twos show up, about 32% chance that one two shows up; and then it decreases for 2 to 5 twos to show up; and for greater than 5 twos to show up there is practically NO chance.

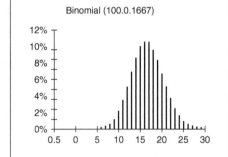

The histogram represents the situation when the die is rolled 100 times and where a two is "success" and anything else is a "failure."

The distribution shapes like a normal distribution, with a mean at 16:67 and a standard deviation of about 3:5

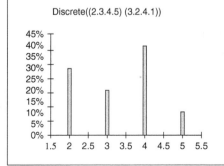

Discrete distribution—each outcome has a value and a probability of occurrrence. The sum of probabilities of occurrences must equal 100%

This distribution may offer flexibility but it does not reflect the fact that in real projects it is rare to find an event which may jump from one value to another. Usually there is a continuous distribution that may follow the trend described by discrete distribution.

- Triangular distribution—It has close properties to pert distribution and, in many cases, is more intuitive.
- Uniform distribution—Also called the "no clue" distribution. Caution, do not be confused by "no clue events," which refer to probability of event occurrence.

Continuous distributions are commonly used in RBE since they better articulate the SME's best judgment. Table 3.8 presents, with comments and examples, the most commonly used continuous distributions.

A risk's range is given by two numbers, the LOW and HIGH, and the shape of a risk's distribution is defined by the position of the *most likely* relative to the LOW or HIGH. The following paragraphs will expand on the range and shape of continuous distributions by analyzing the pert distribution.

Pert Distribution

Pert distributions have a high degree of flexibility, which makes them well suited to model the majority of RBE events. It can take the most shapes of risk distributions used in the RBE process. The information provided on the following pages could be applied very well to all continuous distributions defined by three points. The Pert distribution has more flexibility on adjusting its tails. However, the triangular distribution is more intuitive. It depends on the risk elicitor on which kind of distribution is used. Figures 3.29 through 3.31 present the symmetrical, positive skewed, and negative skewed Pert distributions.

The Pert symmetrical distribution is presented in Figure 3.29. The reader should be reminded about base variability distribution. The most likely value represents the distribution value with the highest chance of occurring. In the case of symmetrical distributions, the most likely value coincides with the distribution's median and mean values. The symmetrical distribution is used when an expert is considering that the values of a risk's impact are as likely to be above as below the most likely value.

When an expert thinks that the risk's impact has a higher density on the lower or higher side of the most likely value, then a Pert asymmetrical distribution can be employed. Figure 3.30 represents a situation where the experts have determined that the lower end of the impact's distribution is dominant. A similar situation may arise when the higher end of the impact's distribution is dominant.

In vernacular terms, the experts think that the risk values are more likely to be at the lower end of the range. It is easy to see that the graph doesn't display values beyond $9 million. A rigorous calculation shows that only 0.3 percent of total possible cases will have a value greater than $8 million. In plain language, only 3 out of 1,000 iterations (plausible cases) will have a value higher than $8 million. So far this is okay; however, what the experts meant when they decided on $10 million for the maximum value may need further attention.

In order to address this issue, the risk elicitor must inform the experts that the maximum value ($10 million) is not reachable. If the expert considers that the maximum value ($10 million) is practical—in a sense, the possibility of having a risk value very close to its maximum—then

TABLE 3.8   Continuous Distributions

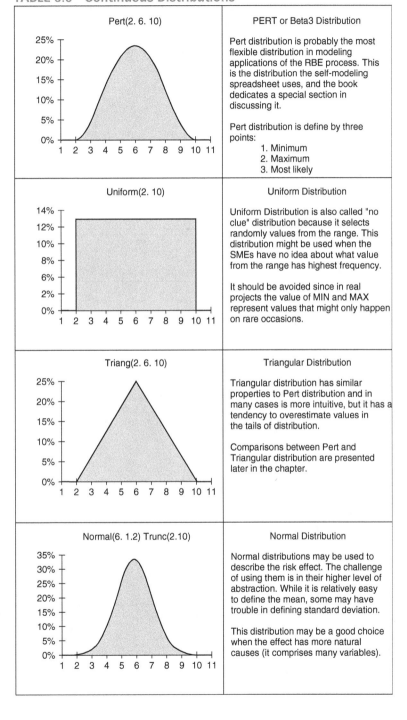

Pert(2. 6. 10)	**PERT or Beta3 Distribution**
	Pert distribution is probably the most flexible distribution in modeling applications of the RBE process. This is the distribution the self-modeling spreadsheet uses, and the book dedicates a special section in discussing it.
	Pert distribution is define by three points: 1. Minimum 2. Maximum 3. Most likely
Uniform(2. 10)	**Uniform Distribution**
	Uniform Distribution is also called "no clue" distribution because it selects randomly values from the range. This distribution might be used when the SMEs have no idea about what value from the range has highest frequency.
	It should be avoided since in real projects the value of MIN and MAX represent values that might only happen on rare occasions.
Triang(2. 6. 10)	**Triangular Distribution**
	Triangular distribution has similar properties to Pert distribution and in many cases is more intuitive, but it has a tendency to overestimate values in the tails of distribution.
	Comparisons between Pert and Triangular distribution are presented later in the chapter.
Normal(6. 1.2) Trunc(2.10)	**Normal Distribution**
	Normal distributions may be used to describe the risk effect. The challenge of using them is in their higher level of abstraction. While it is relatively easy to define the mean, some may have trouble in defining standard deviation.
	This distribution may be a good choice when the effect has more natural causes (it comprises many variables).

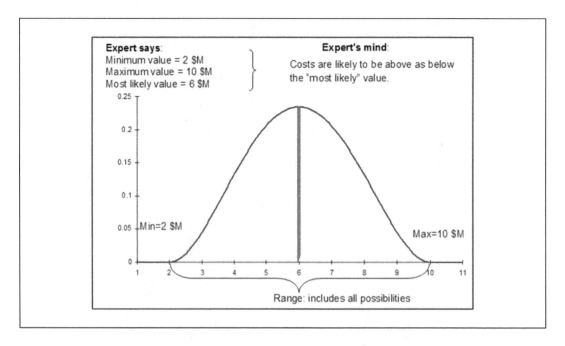

Figure 3.29   Pert Symmetrical Distribution

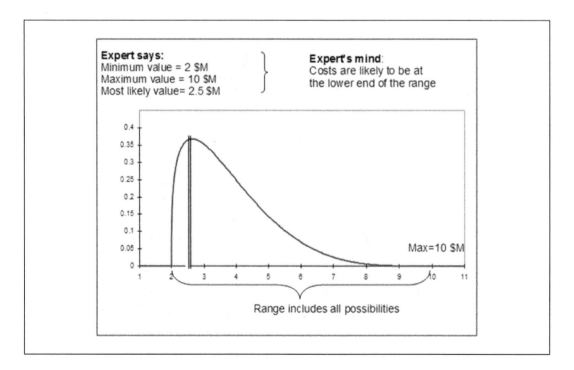

Figure 3.30   Pert Positive Skewed Distribution

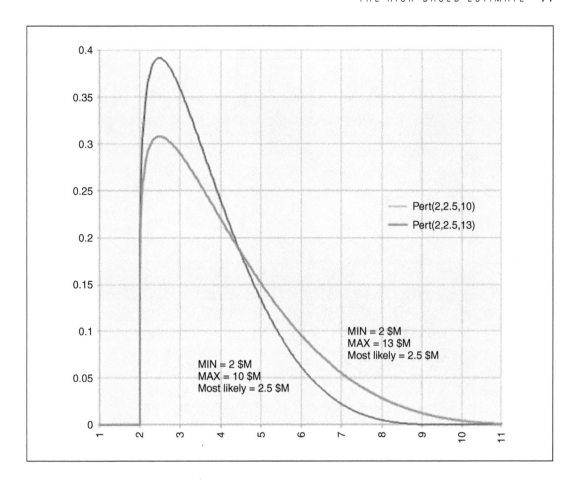

Figure 3.31 Adjusting MAX Value for Better Representation of the Risk's Higher End

the risk elicitor may suggest: (1) increasing the value of the higher end to impractical values, understanding that those high values are unreachable, or (2) using a triangular distribution.

Figure 3.31 presents a situation when the risk elicitor alleviates the condition of not having values in the proximity of the maximum ($10 million) and preserving the requirement that the risk values be concentrated on the lower end of risk range. By increasing the risk's maximum value to $13 million, the model will pick less than 0.2 percent of its plausible cases above $10 million.

The second alternative of replacing the Pert distribution with a triangular one is presented on Figure 3.32. There is a noticeable and dramatic increase of frequencies at the upper end. The Pert and triangular distributions are quite different in the way they represent the upper end. As a matter of fact, they are different, and during a risk's quantification both alternatives should be presented in order to arrive at an informed consensus.

In summary, the issue of having a heavily skewed distribution that represents a risk's impact requires special attention. A visual display of a risk's impact may bring clarity when the experts

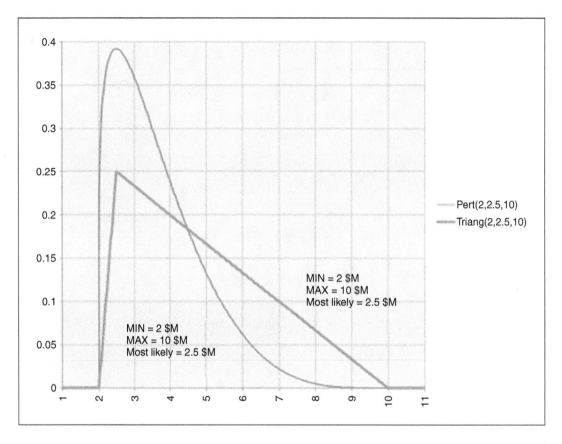

Figure 3.32   Pert Distribution versus Triangular Distribution

can see before their eyes how the risk lays out. There are two major solutions that can be easily displayed: (1) maintain Pert with expended extreme values, or (2) replace the Pert distribution with a triangular one. It is up to the experts to choose the best solution that will fit their vision of how the risk will likely unfold.

Pert — Median versus Most Likely

As presented earlier, the shape of a Pert distribution is given by the position of the third point relative to the range's ends. The third point is called the most likely value and represents the distribution's mode (the value with highest frequency).

In an early stage, the authors assumed that the SMEs were more comfortable with providing the impact median value as the third distribution point. The median value has the property of breaking the distribution into two halves. Half of the plausible cases will be to the left of the median value and the other half will be to the right. This may happen regardless of whether there is a symmetrical or an asymmetrical distribution. In the case of a symmetrical distribution, the distribution's mode coincides with the distribution's median and mean.

Later, based on the observation of thousands of risks elicited, the authors have reached the conclusion that the assumptions we made regarding the distribution's third point were most often incorrect. In most instances, when the risk elicitor has asked for the median value, the SMEs were providing values equal to the MIN or MAX, or values very close to MIN or MAX, which indicated that the SMEs did not look at the third point value as the median value. Based on further study, it was clear that the SMEs were providing the mode value.

Moreover, specialized software that creates simulations requires that when a Pert distribution is employed, the most likely value should be defined. That means that if the risk elicitor goes through the pain of defining a risk's median value, the modeler needs to convert the median value into most likely outside of the model itself.

As a preamble, the risk-based estimate self-modeling (RBES) spreadsheet (presented later in this text), in one of its older versions, allowed the user to enter the "median guess," which was referred to as the "best guess" value and the model then calculated the most likely value. The newer version of RBES requires the most likely value as the third point in order to define the shape of distribution.

We have noticed situations when the median value (best guess) was used as the most likely value. If the best guess is truly the median value of the risk impact and it is used as most likely, then it may create a problem with the analysis. Figure 3.33 shows superimposed the risk impact when the median value is used as a most likely value versus when the calculated most likely value is used. There is a significant change in the distribution's shape and this change is further accentuated if the distribution has a higher degree of asymmetry.

In other words, if the elicited impact median values are used as the distribution's most likely value, all statistical parameters of the elicited impact distributions will change in the model. Table 3.9 shows how the statistical parameters change during this process.

While it is obvious how the change of the mean, mode, and median affect the distribution, it is less obvious how the change in skewness and kurtosis affects the results.

Skewness is a measure of the distribution's symmetry, or more precisely, it is a measure of the distribution's lack of symmetry. A distribution is symmetrical if it looks the same to the left and right of the center point. A higher skewness factor means that there is a higher asymmetry for the distribution and for the risk's impact, and represents a higher frequency at one of the extremes. In many cases, experts want this kind of distribution when they evaluate the value of the third point. So having the most likely value replaced by an elicited median value will thwart the expert's intentions and alter the results.

Kurtosis is a measure of whether the distribution has a peak or is flat relative to a normal distribution. High kurtosis tends to have a distinct peak near the mean, decline rather rapidly, and have heavy tails. Low kurtosis tends to have a flat top near the mean rather than a sharp peak. A uniform distribution would constitute an extreme case of low kurtosis. The kurtosis declines when the most likely value is replaced by the elicited median value. That means that the distribution is flattening which in many cases the experts did not intend.

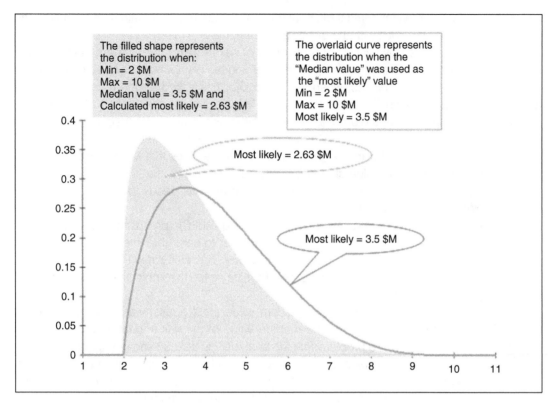

The filled shape represents the distribution when:
Min = 2 $M
Max = 10 $M
Median value = 3.5 $M and
Calculated most likely = 2.63 $M

The overlaid curve represents the distribution when the "Median value" was used as the "most likely" value
Min = 2 $M
Max = 10 $M
Most likely = 3.5 $M

Most likely = 2.63 $M

Most likely = 3.5 $M

Figure 3.33   True Most Likely Value versus Median Elicited Value used as Most Likely

In conclusion, the distribution's most likely value is the value that the expert is most comfortable giving as the third point in the estimate and it is the value that needs to be entered into the model. Eliciting the median values and presenting them as the distribution's most likely values is bad practice. In the worst-case scenario, the elicitor may ask for the median value and make sure that this value is converted properly to the distribution's most likely value.

The best way to elicit risks is to have them displayed in front of the SMEs and to make sure they understand and agree with the range and shape of the risk they are defining. This approach may require additional time for elicitation but it creates a better elicitation environment and builds trust and cooperation among team members.

Triangular Distribution versus Pert Distribution

Pert distributions and triangular distributions have similar characteristics: (1) both are defined by three points—LOW, HIGH, and MOST LIKELY—and (2) both are intuitive. Besides their similarity they are different and a specific risk's impact may prefer one distribution against the other. Figures 3.32 and 3.34 show a significant difference between Pert distributions and triangular distributions when the distributions have high skewness (i.e., the most likely is on one of the

TABLE 3.9 Change of Statistical Parameters When the Median Elicit Value Is Used as the Most Likely Value

Statistical Parameters	Median guess =3.5		Delta
	Used the Median as the Most Likely Value (3.5)	Used the True Most Likely Value (2.63)	
Minimum	2	2	0%
Maximum	10	10	0%
Mean	4.3333	3.7533	15%
Mode	3.5	2.63	33%
Median	4.1392	3.4961	18%
Std. Dev	1.3744	1.2509	10%
Variance	1.8889	1.5646	21%
Skewness	0.6063	0.8981	-32%
Kurtosis	2.8235	3.4087	-17%

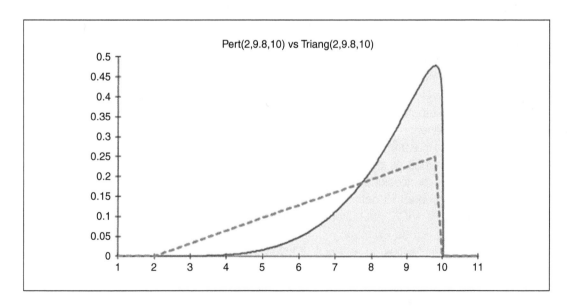

Figure 3.34   Pert versus Triangular with High Skewness (The Most Likely Very Close to MAX)

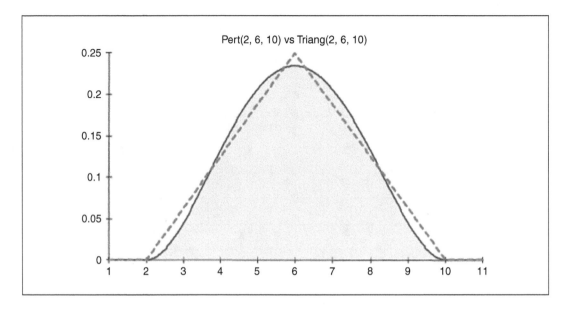

Figure 3.35   Pert Distribution versus Triangular Distribution—Symmetrical

end's proximity). It is recommended that both distributions be displayed so that the experts may decide which one is most appropriate.

In the case of symmetrical distribution, the Pert and triangular will behave in almost the same manner. Figure 3.35 shows that there is no significant difference between these two distributions.

Looking back at the base uncertainty it is important to remember that the variability of base cost and base duration were represented by symmetrical Pert distributions. It is still recommended that Pert distributions represent the impact of base variability despite the fact that triangular distribution is simpler and similar in effect. The reason for this recommendation is the fact that the Pert distribution offers the advantage of smoother tails that may better represent the epistemic component of the base uncertainty.

It has been found in some recent literature the recommendation of using a distribution given by two triangular shapes as presented in Figure 3.36. The distribution is called the *double triangular* and some professionals have expressed preference for it. The double triangular distribution was presented to define the distribution cost of a critical item. It is used in the process presented by AACE International under the name of "Risk Analysis and Contingency Determination Using Range Estimating." The process is different from the process described by RBE with some similarities. In essence, it involves a so-called *perpetual risk* when a risk's probability of occurrence is 100 percent.

If the situation is as described, then considering two events in a mutually exclusive relationship is more appropriate. (There will be more about relations among risks in the following chapters.) We would like to remind readers that a major attribute of RBE consists in providing information to project management that may be used in their efforts of optimizing a project's objectives.

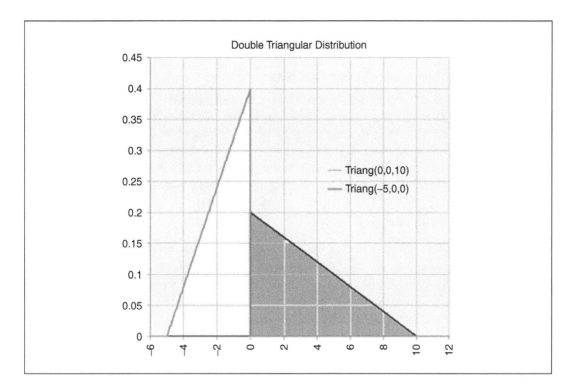

Figure 3.36   Sophisticated Distribution—Double Triangular

The double triangular distribution does not differentiate the conditions when negative or positive values are to be selected.

RBE would treat the double triangular distribution presented in Figure 3.36 as a combination between an opportunity described by 50 percent probability of occurrence and shape given by "triangular with MIN = −5, Most Likely = 0, and MAX = 0" and a threat that occurs only and always when opportunity does not occur with a shape given by "triangular with MIN = 0, Most Likely = 0, and MAX = 10". This new definition of "double triangular distribution" brings data to act upon so management can take proactive measures to improve the project's objectives.

Recommendations

Selecting the Right Distribution

Typically the RBE uses simple distributions that may be easily understood by SMEs, project managers, and stakeholders. The discrete distribution may be used on rare occasions when a specific date represents a constriction to the project or when a certain fixed cost (i.e., no variation allowed) may or may not occur.

The continuous distributions presented previously may be recommended by the risk elicitor for approval by the SMEs. The Pert and triangular distributions are the most common because they are intuitive. Generally, the normal or uniform distribution will be used only when special conditions warrant them.

Using the "LOW, HIGH" Instead of the "MIN, MAX"

The MIN and MAX values are the best choice to be used when the extreme points of a distribution are defined. There are situations when an SME may be reluctant to give a value for the MIN or MAX. In this case, we recommend using the LOW and HIGH terms and identify their meaning. In both cases it is important that the SME understand exactly the meaning and significance of what they estimate. To the extent possible, it is recommended that the distribution shape be displayed in front of the SMEs so that they will understand exactly what they are estimating.

RBE and Monte Carlo Method

The RBE process is illustrated in Figure 3.9. So far we have presented basic information about how base cost and schedule are validated and how risks are identified and quantified in order to employ the Monte Carlo Method (MCM). The following paragraphs present how the collected information on base and risks is processed. At the beginning of this chapter we concluded that the integrated cost and schedule risk analysis provides the most benefit to understanding the project challenges.

The integration of cost and schedule must be seen through the universally accepted concept of the project's triad—scope, schedule, and cost—that provide answers to the followings questions:

- What is the project?—Scope
- How long will it take to complete the project?—Schedule
- How much is it going to cost?—Cost

The triad's elements are interdependent and consequential to the project's cost and schedule estimate. In other words, a change of one element, say, schedule, will affect the cost.

Deterministic estimates consider the triad's dynamics as inherently rigid where the schedule and cost elements converge to a single point. This convergence point is expanded into an uncertainty zone by contingencies added to cost and/or schedule. The direction of the uncertainty zone is only upwards since contingencies are added on top of the deterministic estimate (Figure 3.37).

(Note: For better understanding of the visuals in this section, we consider the magnitude of a triad's element to be proportional with its length.)

The uncertainty zone presented within the deterministic concept is, essentially, a blank check to deal with anything that may happen. The value of the check is subjective to the estimator's experience.

Risk-based estimates are not without checks, but the process of risk analysis clarifies the purpose of the check. Further, risk analysis enables risk management and thereby influences the number of zeros on the checks.

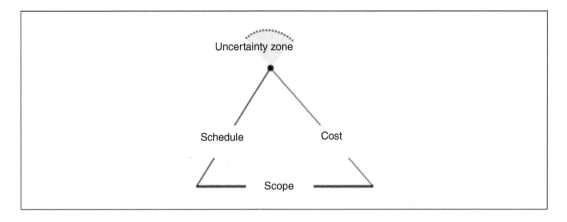

Figure 3.37   Deterministic Triad—Scope, Cost, and Schedule

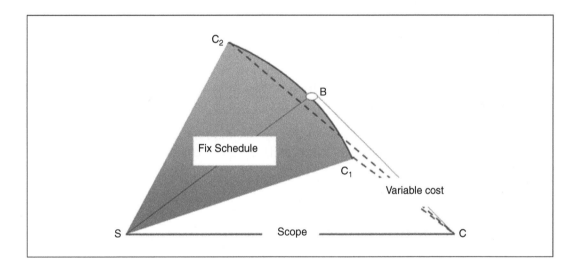

Figure 3.38   Cost Risk–Only Triad

In the remaining section we will look at possible scenarios of project risk analysis with fixed scope.

1. *The project must be delivered at a fixed date* or only the cost of the project may change. This scenario represents the typical situation of cost-only risk analysis. Any event that may delay the project's completion date is terminated through added cost. Figure 3.38 illustrates the project's cost and schedule triad dynamics when the delivery date is immovable.

Note: The cost and schedule segments connector (point B) resides on the arch with the center on S. The fixed schedule is represented by the arch C_1-B-C_2 and the cost range is represented by the length of segments CC_1 and CC_2. The cost opportunity zone (events that may reduce the project cost) are points within the BC_1 arch, whereas the cost threat zone (events that may increase the project cost) are points within the BC_2 arch.

2. *The project must be delivered at a fixed cost* or only the schedule of the project may change. This scenario represents the typical situation of schedule-only risk analysis. Figure 3.39 illustrates the project's cost and schedule triad dynamics when the cost is unchangeable.

Note: The cost and schedule segments connector (point B) resides on the arch with the center on C. The fixed cost is represented by the arch S_1-B-S_2 and the schedule range is represented by the length of segments SS_1 and SS_2. The schedule opportunity zone (events that may shorten the project schedule) are points within the BS_1 arch, whereas the schedule threat zone (events that may delay the project schedule) are points within the BC_2 arch.

3. *Project has both cost and schedule flexible* or both cost and schedule may change at the same time. In this case the cost and schedule are integrated in a comprehensive risk analysis and the project's triad looks like that shown in Figure 3.40. The triad has two degrees of flexibility and the cost and schedule uncertainties define the so-called *uncertainty cloud*.

The uncertainty cloud has four distinct zones: (1) cost and schedule opportunities where the project is delivered earlier than planned and under budget; (2) cost and schedule threats where

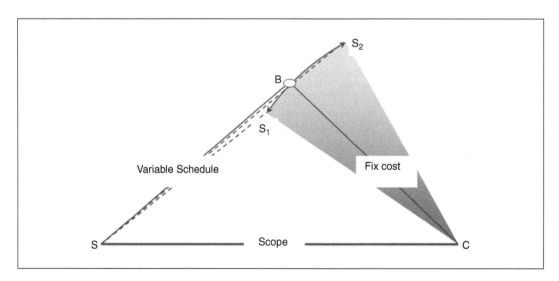

Figure 3.39   Schedule Risk–Only Triad

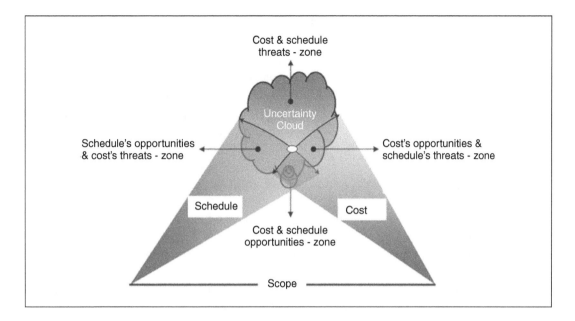

Figure 3.40   The Risk Cloud Triad

the project is late and experiences cost overruns; (3) cost opportunities and schedule threats where the project cost is less but it takes longer to build; and (4) schedule opportunities and cost threats where the project is delivered earlier but the cost is more than the deterministic estimate.

Cost and Schedule Estimates

The previous paragraphs presented a theoretical explanation of how risk analysis may be conducted using the project's triad paradigm. The reality of project management shows that it is equally important to estimate both the cost and schedule of a project. One good reason why a cost risk analysis should integrate cost and schedule is the simple fact that they are inseparable and the integration provides better and richer data. Experience has demonstrated that a delay in the project schedule constitutes an indirect increase to the project cost. The old saying "time is money" has greater value than many people think. At the project level there are resources (rented equipment, labor, project support personnel, and so on) that are very sensitive to project delay. If the delay is significant, then inflation may increase the cost without bringing any additional value to the project.

The RBE recognizes the undividable value of the project's triad (scope, cost, and schedule) and the necessity of treating the cost and schedule together once the scope is defined. The RBE may have two options: (1) analyzing the risk of the project cost and schedule separately, as shown in Figures 3.38 and 3.39, and (2) analyzing the risk of the project cost and schedule simultaneously. The first option is called *nonintegrated* cost and schedule risk analysis and the second option is called *integrated* cost and schedule risk analysis. Each option has its own advantages and disadvantages and may be used at the discretion of the project manager.

Nonintegrated Approach of Cost and Schedule Risk Analysis

The nonintegrated approach of a project's cost and schedule risk analysis is limited in the type of data that it can provide and can be performed without the contribution of an experienced risk modeler. The analysis is performed for the cost and schedule separately. Figures 3.2, 3.38, and 3.39 represent the essence of the nonintegrated approach of RBE. After the distributions of the cost and the schedule are determined, the estimator and scheduler may decide how to articulate their combined effect. By doing this, the method introduces undesirable subjectivities into the final results.

Since the scheduler and the estimator together with risk elicitor and other SMEs must go to the effort of identifying and quantifying risks for the cost and schedule, why not take the additional step of integrating them? The answer is simple: The integration of cost and schedule requires advanced knowledge of modeling. Knowledge of modeling is expensive. There is software available that may facilitate the integration of the cost and schedule, however, it still requires a trained risk modeler to develop and populate the model correctly. The risk-based estimate self-modeling tool that we will present later assuages this requirement by providing a template ready to accept the risk data with minimum knowledge of risk analysis.

Integrated Approach of Cost and Schedule Estimate

The integrated approach to project cost and schedule risk analysis leads to better results because it binds together cost and schedule for every single situation. Each plausible case is defined by an algorithm and random variables of the cost and schedule elements (i.e., base estimate, risks, inflation, and so forth). This integration provides for a robust and comprehensive understanding of the project's prospects.

The project flowchart constitutes the basic algorithm of the model. The flowchart shows only critical activities that may dictate the delivery of the project and it is an abstract of the existing detailed project schedule. In the next paragraphs we will expand on the project flowchart.

Project Flowcharts

The integration between cost and schedule is facilitated by the creation of a project flowchart. The flowchart has to be easy to understand and should have a limited number of activities. Years of experience examining hundreds of flowcharts of various projects (a variety of sizes, complexity, different levels of design and requirements) and their influence on the analysis results has once again led the authors to embrace the KISS principle.

For example, the sophistication of the flowchart displayed on Figure 3.41 brings no value to the analysis or to the understanding of the project. The majority of the activities are not affected by any events that might change the outcome of the analysis. Having many activities introduces a dispersion of the base cost with implications related to the number of variables that

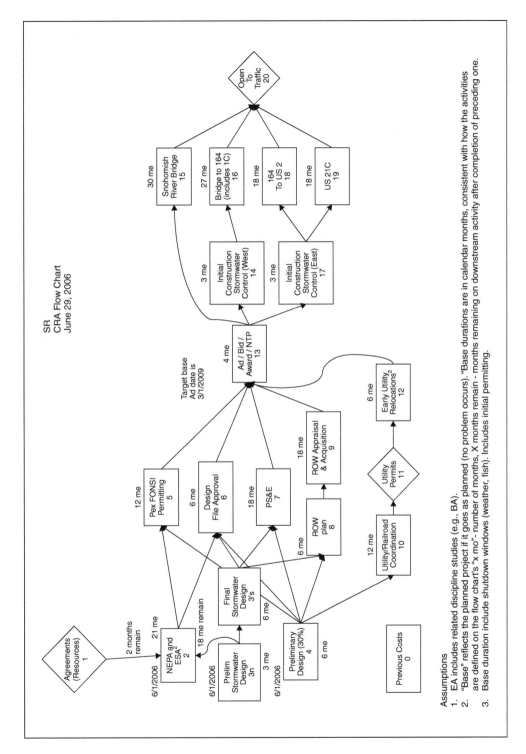

Figure 3.41 Design-Bid-Build Flowchart—The Sophisticated Approach

an analysis has. If the flowchart is unnecessarily complicated, the model will be more complex and the danger of introducing errors increases. At the same time, the SMEs may lose interest in discussions because of difficulty in following the project's flow. Having a jaded SME is the most dangerous effect of any unreasonable complication.

The definition of a project's flowchart used on risk-based estimate is: The project's flowchart diagram used on the risk-based estimate (cost risk analysis) is the project's critical path (CP) schedule at the level of significance at the time of estimate. This definition has two major statements: (1) the flowchart diagram must present the critical path schedule, and (2) the flowchart diagram must be synchronized with the quality of data used in the estimate.

The first statement indicates the importance of having a CP schedule representation on the project's flowchart. At the same time it is important to understand that the flowchart represents the base CP schedule. There may be situations when the critical path changes under certain schedule risks. In such a case, the risk elicitor must identify the risk impact through the new path. In this case, having a detailed project CP schedule may be useful. The ultimate objective of assessing the schedule risk impact is identifying its effect on the last activity of the base CP schedule.

Usually, it is recommended to limit the number of activities included in the flowchart to the absolute minimum necessary. For example, all projects that follow the process of design-bid-build are well represented by the flowchart presented in Figure 3.42. The flowchart brings

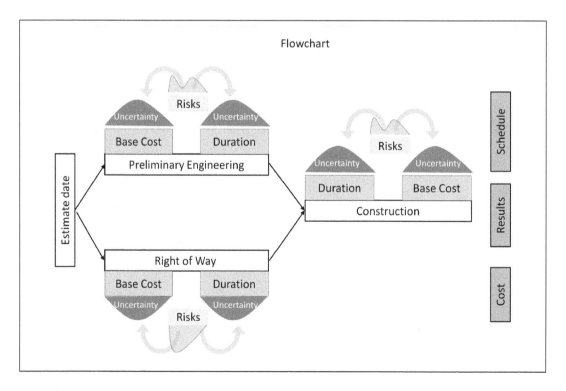

Figure 3.42   Design-Bid-Build Flowchart—The KISS Approach

together the values quantified for the base estimate (cost and schedule) and risks (cost and schedule) and creates the algorithm that the MCM will employ in order to finalize the statistical analysis as described by Figure 3.9.

Figure 3.42 depicts the project algorithm—how the project is going to be delivered from the time of the estimate. It must be remembered that an estimate is nothing more than a snapshot in time. So the time of the estimate is the beginning of the flowchart. The *estimate date* has double meaning: (1) it specifies the date of market prices and (2) it's the starting point of computing the impact of inflation. In other words, the estimate date implies an expectation that all values introduced in the analysis are the best ones available at that time and is represented in terms of present net value.

The preliminary engineering (PE) or design phase includes all of the necessary tasks to be performed in order to have the project ready for construction. This activity is assigned a base cost and base duration with the associated variability.

Similar data is assigned to the right of way (ROW) activity. The ROW is a term taken from highway projects and it is related to the right of use of the land (i.e., real estate). The ROW activity must finish before construction starts.

The ad-bid-award is a short activity that includes the efforts of advertisement, acceptance and validation of bids, and bid award. This activity has a duration dictated by the project owner's policy. The duration of this activity includes the time from when the advertisement was initiated to the time when the construction begins.

Construction (CN) is the dominant activity of the flowchart diagram. Usually it is shorter than the other two but is more expensive. This activity has assigned to it a base cost with its uncertainty (variability plus market condition) and duration with its variability.

Figure 3.42 shows that each activity is affected by risks. A risk may affect the activity's cost, duration, or both. During the simulation process for each of iterations (plausible case), a risk may or may not affect the activity depending on how the random number comes up and, when it occurs, its impact is a value taken from the risk distribution.

Monte Carlo Method Applied to RBE

Figure 3.42 and Table 3.10 show how the Monte Carlo method is applied to RBE. For a plausible situation, the model extracts a random value for the base cost of the preliminary engineering (PE), right of way (ROW), and construction (CN) according to their cost distribution.

The model then tests each risk as applied to each flowchart activity. If a risk was meant to occur (the game of random numbers) the model will extract a quasi-random value from risk's distribution that describes its impact. All of these values are then added together and form the total cost of the project in current year (CY) dollars.

Table 3.10 shows a micro-project (for illustration only) that has all three activities presented in Figure 3.42. The estimate is performed only for current year dollar so it doesn't include the activities' duration as presented in Figure 3.42. In other words, the micro-project is an example of nonintegrated cost risk analysis. The preliminary engineering, right of way, and construction costs are defined by their range (preliminary engineering: LOW = 2.5 and HIGH = 3.5; right of way: LOW = 11 and HIGH = 13; and construction: LOW = 32 and HIGH = 35)

TABLE 3.10 How the Monte Carlo Method Works

Phase	PE	ROW	CN	PE	ROW	CN	Total Cost
Range	2.5 to 3.5	11 to 13	32 to 35	.5 to .8	1 to 5	2 to 8	
Percentage	100%	100%	100%	50%	30%	67%	
Iteration #	Base Cost			Risks Cost when Occurs			
1	3	11.6	33	0.7	4	6	58.3
2	2.9	11.9	34	0.55		3	52.35
3	2.8	11.3	33.5				47.6
4	3.4	12	32.9	0.65	3		51.95
5	3.2	12.3	34.6	0.63		7	57.73
6	3.3	12.7	33.5		2		51.5
7	2.8	11.8	33.8	0.66		5	54.06
8	2.6	12.4	32.9			6	53.9
9	3	11.7	34.3			7	56
10	3.3	12.6	34.2	0.76		4	54.86
11	2.9	12.4	33.4		4		52.7
12	2.9	12	33.8	0.65		5	54.35

Total cost	Min value	47.6
Total cost	Max value	58.3
Total cost	Median value	54.0

and 100 percent probability of occurrence. The 100 percent probability of occurrence means that during the simulation a value from the respective range will always be extracted and used in the calculations. The base estimate (cost or duration) will always be represented in each plausible case.

The risk side is a different story. A risk is characterized by its probability of occurrence that will dictate when the risk would occur. In the micro-project example, preliminary engineering is affected by a risk that has a 50 percent chance of occurring and its impact range is (LOW = 0.5 to HIGH = 0.8) units. The ROW activity is affected by a risk with a lower probability of occurrence (30 percent) and the range of its impact is from 1 to 5 units. Finally, the construction activity is affected by the most significant risk that may occur (66 percent) and has an impact of 2 to 8 units.

Table 3.10 indicates how the MCM works. Each row below the header represents one plausible case and in modeling terms is called *iteration* or *realization*. The table shows only 12 rows but a real model typically runs thousands of plausible cases.

It is noticeable that all cells under the "Base Cost" are filled with numbers and the cells under "Risks Cost When Occurs" are mixed (some are filled, some are empty). That means that for every plausible case the model always picks a number from the base and only a random from the risk impact.

The number that is picked from each distribution depends on the random number and the type of distribution. The micro-project example assumes that each distribution is uniform. So each number from the distribution range should have an equal chance of being selected. That is why the uniform distribution is called the *no clue* distribution. The distribution doesn't have any "preference."

After all iterations are completed the model creates a database that will be used on computing the values of interest. Table 3.10 shows one value of interest, which is the *total cost,* but the following paragraphs will present the typical results associated with the RBE.

The preceding description is simplistic, but it has the benefit of being intuitive and easy to understand. Reality is more complex and the model must consider all possible situations such as: risks' dependency, risks' correlation, inflation factors, and the schedule dynamics in order to define current year (CY) and year of expenditure (YOE) cost estimates, and the "end of construction date."

THE MICRO-PROJECT

The risk-based estimate may provide results in different forms: (1) tables, (2) histograms and cumulative distribution functions, (3) tornado diagrams, and (4) risk maps. The last two focus on identifying the most critical risks among the significant ones to assist with risk management efforts. The percentile table gives information about what chance a project has to overrun a certain budget.

Percentiles Table

For example, the micro-project discussed in the preceding paragraphs has the percentiles table presented in Table 3.11. The unit used on the micro-project is millions of dollars.

Table 3.10 provides useful information to both project manager and risk analyst. The reader is going to be interested in the chances of staying within budget. For example, the micro-project is approved for a budget of $54 million. According to the cost risk analysis just performed, the project has about a 60 percent chance of staying within budget.

Using the vernacular, it would be stated that there is a 60 percent confidence level of delivering the project below the budget figure. In other words, a 60 percent confidence level shows that there is about a 40 percent chance that the project will overrun the budget. Since there is a significant chance of overrunning the budget the project manager (PM) wants to know what he or she can do to reduce the probability of this happening. In this case the PM has an easy task because the project has only one risk on each activity.

In many cases the PM looks at the preliminary engineering percentiles, right of way percentiles, and construction percentiles and analyzes the chances of overrunning the phase budget. At

TABLE 3.11   Micro-Project Percentiles

Name	Preliminary Engineering (PE)	Right of Way (ROW)	Construction (CN)	Total Simulated Cost (TSC)	Sum of PE, ROW, and CN (SUM)	Delta (TSC minus SUM)	Delta % (Delta/TSC)
Minimum	$2.50	$11.00	$32.00	$45.80	$45.50	$0.30	0.65%
Maximum	$4.29	$17.96	$42.90	$63.42	$65.14	-$1.72	-2.72%
Mean	$3.32	$12.89	$36.79	$53.00	$53.00	$0.00	0.00%
Mode	$3.40	$12.80	$34.89	$52.68	$51.09	$1.59	3.02%
5%	$2.60	$11.15	$32.43	$47.82	$46.18	$1.64	3.44%
10%	$2.71	$11.29	$32.90	$48.56	$46.89	$1.67	3.44%
15%	$2.81	$11.43	$33.34	$49.11	$47.58	$1.53	3.12%
20%	$2.90	$11.57	$33.75	$49.69	$48.22	$1.47	2.96%
25%	$3.00	$11.71	$34.15	$50.25	$48.86	$1.38	2.76%
30%	$3.09	$11.86	$34.56	$50.85	$49.51	$1.35	2.65%
35%	$3.15	$12.00	$34.92	$51.48	$50.07	$1.41	2.74%
40%	$3.21	$12.14	$35.81	$52.02	$51.17	$0.86	1.65%
45%	$3.27	$12.28	$36.46	$52.52	$52.01	$0.51	0.98%
50%	$3.32	$12.42	$36.95	$53.02	$52.69	$0.33	0.63%
55%	$3.37	$12.55	$37.45	$53.51	$53.36	$0.15	0.28%
60%	$3.42	$12.68	$37.87	$54.03	$53.97	$0.06	0.11%
65%	$3.47	$12.81	$38.32	$54.50	$54.59	-$0.10	-0.18%
70%	$3.54	$12.94	$38.78	$54.96	$55.26	-$0.30	-0.54%
75%	$3.64	$13.55	$39.23	$55.44	$56.43	-$0.99	-1.78%
80%	$3.74	$14.27	$39.68	$56.00	$57.68	-$1.68	-3.00%
85%	$3.84	$14.94	$40.13	$56.59	$58.91	-$2.33	-4.11%
90%	$3.95	$15.64	$40.64	$57.36	$60.23	-$2.87	-5.00%
95%	$4.06	$16.36	$41.33	$58.50	$61.74	-$3.24	-5.54%

this time the PM initiates the risk response plan examining the most important risks. This will be discussed in detail in subsequent chapters.

Another issue that the PM may observe is the fact that for a certain confidence level the sum of PE, ROW, and CN doesn't amount to the "total simulated cost." The layperson expects these two totals to match. In many situations the eyebrows have been raised and questions have been asked about the validity of the numbers and about what number should be used. The main complaint has been about the fact that the sum of components (PE, ROW, and CN) was significantly higher than the simulated sum, which is true for the most part.

The micro-project has a total simulated cost (TSC) that is 5 percent lower than the sum of its components (SUM) at a 90 percent confidence level. The PM will ask for the 90 percent confidence level to be assigned the $60.2 million, which represents the SUM. Five percent is significant and the discrepancy needs to be explained.

Statistics have many laws and one of them is called *central limit theorem* and may be translated simply as "the sum of means equals the mean of sum." In other words, the summation of the mean value of many distributions is equal to the mean of the distribution created by adding them statistically.

The summation of any of the same confidence levels of many distributions in most cases is not equal to the same confidence level of the distribution created by adding them. There are special conditions regarding the relationship among distributions when the equality of the sum of percentiles equals the percentiles of the sum. This will be discussed later on and these conditions are called "correlation between distributions."

Table 3.11 shows that the "Delta: TSC-SUM" is zero only at the mean value and it changes its polarity from positive to negative at the mean value. This kind of behavior of the newly created column called "Delta" is normal and the risk analyst should examine it and make sure that the analysis makes sense. The delta values are decreasing from positive to negative values while the confidence level increases and the polarity changes at the mean point.

A short digression is appropriate. We would like to cite Sam Savage, a Stanford University professor, who calls this behavior the "weak form of the flaw of averages." Sam states that the combined average of two uncertain quantities equals the sum of the individual averages, but the shape of the combined uncertainty can be very different from the individual shapes[15]. This is correct and we would like to explain in more detail the shape issue [the sum of individual shapes (SUM) and the shape of the sum (TSC)] and provide some guidance about what a risk analyst should expect.

If the threats are dominant, the mean point is above the median and when the opportunities are dominant the mean point is below the median. The analyst should check the relative position of mean value related to the median value keeping in mind what the dominant risks are, threats or opportunities.

Figure 3.43 shows the envelope effect of the SUM over the TSC. The TSC curve is between mean and SUM for the entire range and the graph has only a triple intersection point.

Histograms and Cumulative Distribution Functions

The percentiles table provides numerical information concerning the project cost and schedule. The histograms and cumulative distribution functions (CDFs) provide quantitative information on the project cost and schedule in visual format. These graphs provide "at a glance" information on the project and may help to communicate the project's challenges. Figure 3.44 presents the values of interest of the micro-project in the form of histograms and CDFs. Each graph has

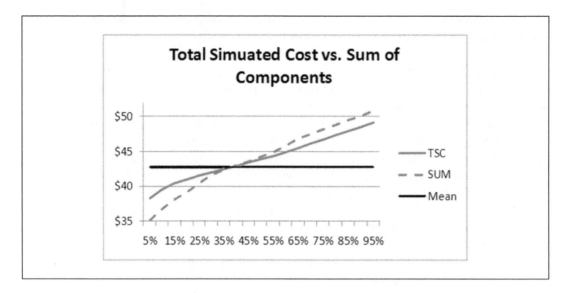

Figure 3.43   Graphical Form of Central Limit Theorem Applied to RBE

its own story to tell. The preliminary engineering graph indicates three distinct sections: (1) the base uncertainty only "from 2.5 to 3," (2) a combination of base uncertainty alone and base uncertainty and risk "from 3 to 3.5," and (3) base uncertainty and risk "from 3.5 to 4.3." It is interesting to observe that the shape of the distribution at the high end (4 to 4.3) of the histogram is gradually tapered to zero and the CDF is reducing its slope.

The right of way distribution is telling the same story but with different nuances. Since the probability of risk occurrence is just 30 percent the transition between section (2) and section (3) is more abrupt and significant. A different story is displayed by the total cost distribution. The shape is moving toward a normal distribution and the CDF is getting its "S" shape.

The analysis of the micro-project stops here since the project is affected by only one risk for each phase and the risk response strategy is quite simple. Analyze each risk and decide what benefits it may bring if they are avoided, mitigated, transferred, or just simply accepted. The next few paragraphs present high-level analysis of a complex project in order to demonstrate the flexibility of RBE. This analysis will be expanded on in Chapter 6, when the RBES spreadsheet is presented in detail.

EASTSIDE CORRIDOR PROGRAM

The RBE is one of the most versatile approaches of quantitative risk analysis that allows users to be creative and innovative in meeting projects' specifics and owners' requirements. The next few paragraphs present the algorithm of an unconventional construction program that consists of short-term objectives and long-term goals. In-depth discussion of the real analysis of this program, including real results, is presented at the end of the Chapter 6.

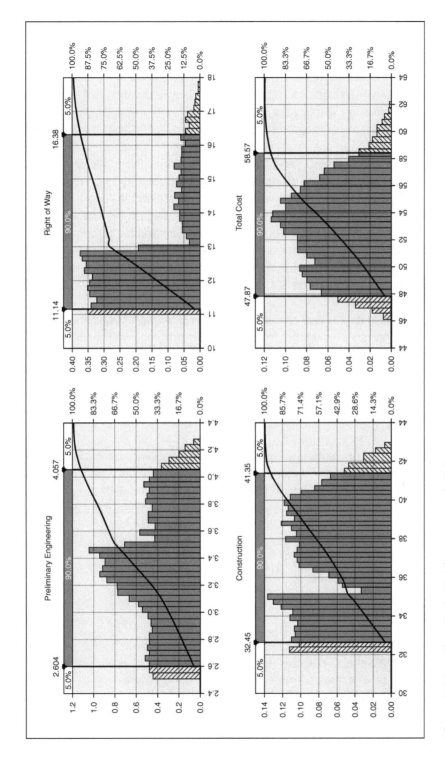

Figure 3.44 Micro-Project's Histograms and CDFs

The program is called "Eastside Corridor" and it has four main construction deliverables with an approximate cost of $1.3 billion. The construction deliverables are called section A through section D and are scheduled in sequence. Figure 3.45 presents the program schedule.

Figure 3.45 clearly shows the algorithm of how the program's tasks are sequenced. Section A is the first segment that will be constructed and has the longer duration but it is not the most costly. Section A construction is planned to be finalized by the end of year 2020 and the other sections are planned to be finalized by the end of year 2022, 2025, and mid-year 2029, respectively.

Section B through section D are going to be delivered far in the future and the quality of data used for their cost and schedule estimates is not yet satisfactory to perform a robust risk analysis. So the RBE process will focus on risks that may affect Section A.

Sections B, C, and D will have assigned uncertainties based on the overall knowledge of each of them. Risks that may affect these sections will be identified for further analyses but the workshop will not quantify them. The owners have indicated special interest on the risk analysis of section A and have recognized that the time and resources that would need to be spent for an in-depth risk analysis of sections B through D will bring little to no value to their current understanding of the program.

In order to accommodate the lack of data of sections B through D and the owners' desire to forgo an in-depth analysis at this time, the RBE for this program was divided into two main components: (1) project level—section A alone, and (2) program level—the entire corridor, which includes all tasks needed to deliver the program. The project level is a component of the program level.

Project Level Analysis

The section A flowchart is presented in Figure 3.46. It is a regular risk-loaded flowchart and is similar to the chart presented in Figure 3.42. Section A is treated like any other project according to the methodology presented in this chapter. Section A results are presented in the form of graphs and tables as shown in Figure 3.47. (More detailed data are presented in Chapter 6.)

For section A, the analysis presents the list of candidates for risk response in the form of a tornado diagram and risk map.

Candidates for Mitigation

A typical *candidates for mitigation* (tornado) graph is presented in Figure 3.48. The tornado diagram gives viewers an immediate image of how the risks are ranked based on their expected impact. Each bar of the tornado diagram represents the product of risk's mean value and its probability of occurrence. The threats are directed toward the right since they are increasing the project cost or duration and the opportunities are directed toward the left since they are reducing the project cost or duration.

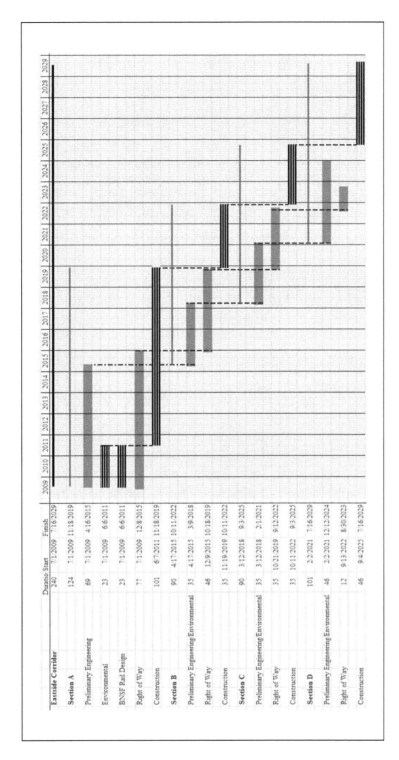

Figure 3.45 Program Level—Schedule

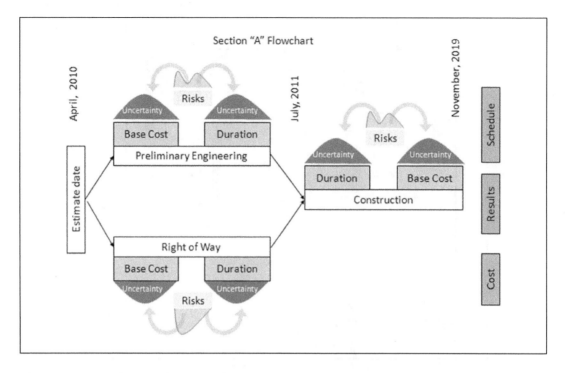

Figure 3.46   Section A—Flowchart Diagram

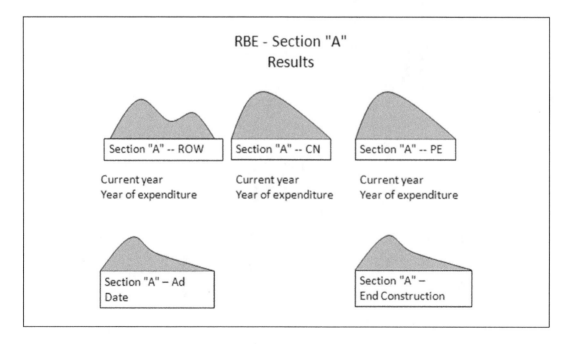

Figure 3.47   Section A—Results Example

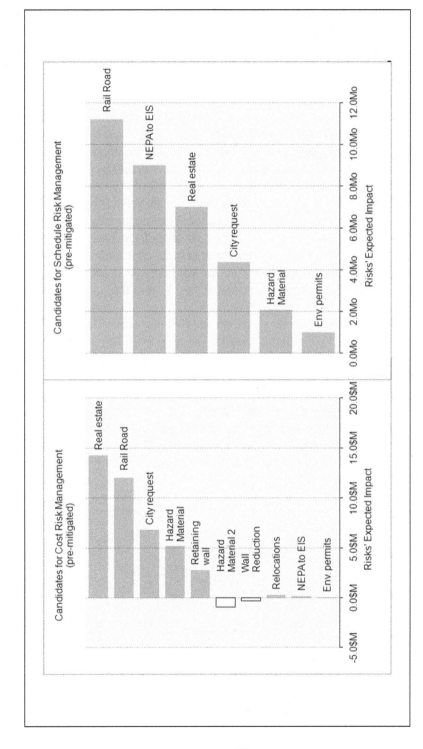

Figure 3.48 Pre-mitigated Section A—Tornado Diagram

It is important to differentiate threats from opportunities because the risk response for each of them is different. The rule is to "minimize or avoid the threats and maximize the opportunities." The risk's characteristics that may be affected by the response strategy are: probability of occurrence and impact. Action may be taken to reduce or eliminate the threats' probability of occurrence or to increase the opportunities' probability of occurrence, while reducing the threats' impact or increasing the opportunities' impact.

The tornado diagram provides useful information about the risk's average magnitude but it may mislead the reader on risks that possess a high impact and a low or very low probability of occurrence. The very low probability and very high impact risks are more dangerous than the high probability and moderate impact risks since the manager may easily ignore them, however, when they do occur the impact is often dramatic.

The risk matrix presented in Figure 3.49 alleviates this deficiency by displaying a risk in 5 × 5 array where the vertical axis represents the probability of occurrence and horizontal axis represents the impact. From left to right the impact is scaled as: Very Low (VL), Low (L), Moderate (M), High (H), and Very High (VH) for the impact as cost and/or duration. From the bottom up the probability of occurrence is scaled as: Very Low (VL), Low (L), Moderate (M), High (H), and Very High (VH). The cost risk is shown by "$" and the schedule risk is shown by "Mo."

The shading suggests different levels of priority for risk response. The shading closer to the upper-right corner represents the first priority for risk response. The shading closer to the lower-left corner represents the last priority for the risk response. Figure 3.49 shows that the risk requires immediate attention because of its cost component. Of course its schedule component will be affected too when the risk is mitigated.

A different way of displaying the risk response priority is illustrated by the project's risks map presented in Figure 3.50. The risks map brings together the significant project risks representing both components: cost and schedule. The low probability and high impact risks are better represented so the project manager may be made more aware of what the future has in store for the project.

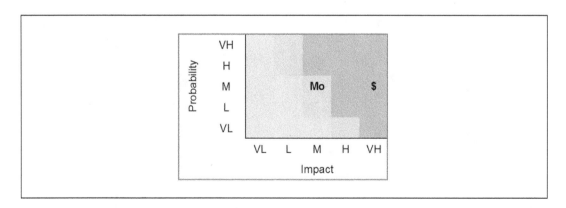

Figure 3.49   Risk Matrix

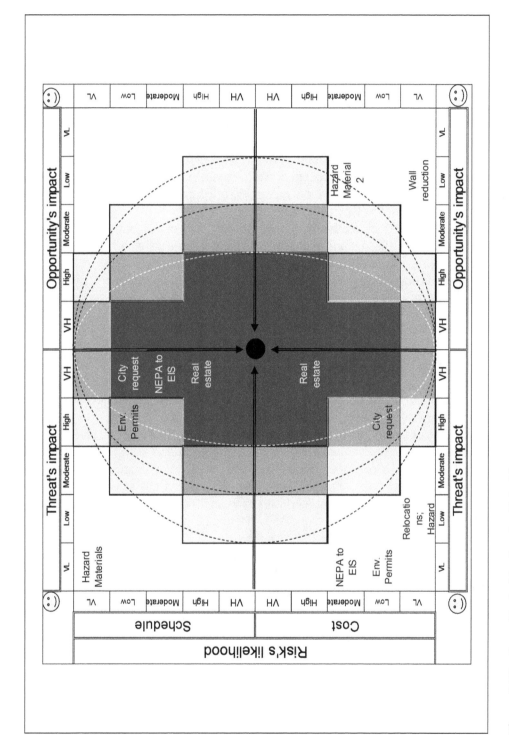

Figure 3.50 Pre-mitigated Section A—Risks Map

The project risks map has four quadrants in order to differentiate threats and opportunities as well as cost and schedule. As a general rule it is recommended that the risk response strategies should be applied first to risks located in the darker zone. In addition to the shading code the broken line ellipses complement the order of importance of risks included in each perimeter.

The shading code and ellipses emphasize the secondary recommendation: the risk impact is more critical than the risk probability of occurrence. The secondary recommendation makes the broken lines curve to become ellipses and not circles, as someone may expect.

Since the risk's impact is more important, in making a decision to respond to that risk or to accept it, than the risk's probability of occurrence the ellipses' short axis are horizontal. In this way risks with high impact values and a low probability of occurrence will be recognized and perhaps will receive the attention they deserve.

For example, Figure 3.50 shows that "City request" has very low probability of occurrence with a very high impact for the cost and schedule. The project manager should pay attention to this risk since if it happens the project will be significantly affected.

The risk analysis is developed to support the project manager's decision on how the project is delivered. The results of risk analysis should be carefully evaluated and understood by the decision maker before a decision is taken. RBE provides to project manager data in different forms in order to facilitate the communication of what really matters.

Program Level Analysis

After Section A is finalized, the program level analysis will begin by supplementing the data entered at Section A with information related to each of the additional sections. It is noticeable that the critical path presented in Figure 3.45 is given by the construction schedule so the assumption made is that all construction activities have a start-to-finish relationship.

Figures 3.51 and 3.52 show the program level model algorithm in two different views. Figure 3.51 shows the basic sections' data and how the sections are set up in the model. Figure 3.52 shows the uncertainty associated with each section (cost and schedule) and the type of inputs of each section. Each section has a different range of uncertainty because they are significantly apart from each other.

For example, the Section B, scheduled to be delivered by the end of year 2022, has a range of −20 percent; +50 percent while Section D, scheduled to be delivered by the middle of year 2029 has a range of −30 percent; +100 percent. The wide range of each section is necessary to accommodate the lack of knowledge and to cover the impact of all risks that may occur.

Furthermore, Figure 3.52 shows that preliminary engineering is calculated as a percentage of the construction cost. As such, the uncertainty of the preliminary engineering estimate mirrors the uncertainty of the construction estimate.

The program level estimate will provide results on the cost and schedule estimate but will not attempt to present any information on program level risks.

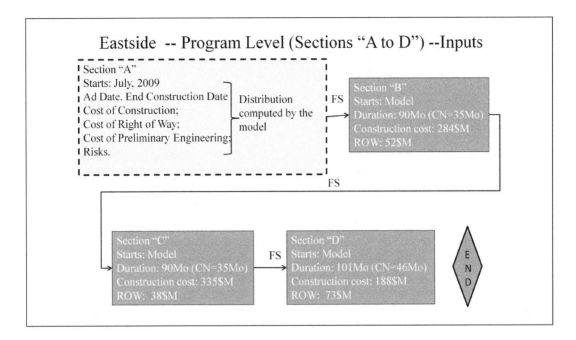

Figure 3.51   Eastside Program Level—Flowchart

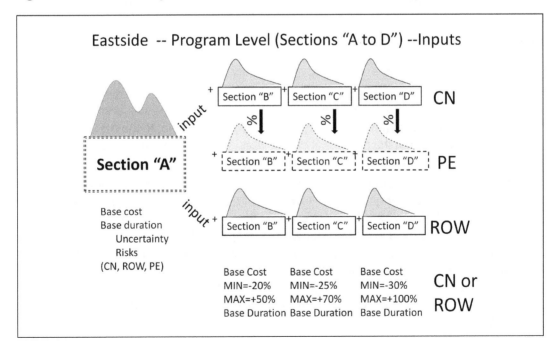

Figure 3.52   Eastside Program Level—Variables

In summary, the Eastside Corridor Program provides an example that shows the flexibility and capabilities of the RBE. The RBE limits depend on risk analysts' willingness to go the extra mile in the pursuit of creating a dependable cost estimate and risk analysis. We remind readers that detailed information about these two projects are presented in Chapter 6.

RBE AND ESTIMATE ACCURACY

It seems that the notion of an estimate's range given by RBE creates confusion among some professionals. It appears that the range magnitude of cost estimate given by cost risk analysis is somehow related to the estimate's accuracy.

Some professionals consider that if the estimate given by RBE has a large range, then its accuracy is low and vice versa. This association is wrong and the following paragraphs will demonstrate why this is so. The issue of understanding the accuracy of an estimate in context of how RBE presents its results (range and shape) was raised by one experienced program manager (PM) who supervised large construction programs.

The PM sent an email with his concerns about his interpretation on issues of the estimate's accuracy. A copy of the content of his email is presented in Figure 3.53. The next few paragraphs present the response that was provided. The dialog presented next allows the reader to capture the authors' position on "the range of the estimate given by RBE" and the estimate's accuracy.

The PM's statements are presented using italics and the authors' responses are in bold.

I'm struggling to understand the definition of "estimate accuracy" and how to calculate it. Here's stuff I'm copying from AACE and my conclusions:

- *Risk is not a measure of estimate accuracy*—**Okay**
- *Each critical item has possible extreme values—a range*—**Okay. As a matter of fact *each item* has a range associated with its cost or duration, but the risk analysis will count only the critical items (items that may have a *significant impact* to the cost or schedule). We recommend that the number of critical items should be no more than 15.**
- *The range is not the accuracy of each item*—**We are not sure what this means. The range and accuracy are not directly connected. It looks as though some professionals associate a "wide range" with a "low level of accuracy" and a "narrow range" with a "high level of accuracy." In our opinion, this association is wrong. It may be true of the opposite. It is possible that a wide range of the estimate is more accurate than a narrow range of the same estimate. The accuracy of the estimate depends on the quality of the data that drives the estimate. The quality of data depends on many other things but that is a different story. Perhaps the notion of *error* should be introduced. High accuracy means smaller and fewer errors; low accuracy means more and larger errors.**

I'm struggling to understand the definition of estimate accuracy and how to calculate it. Here's stuff I'm copying from AACE and my conclusions:

- Risk is not a measure of estimate accuracy
- Each critical item has possible extreme values—a range
- The range is not the accuracy of each item
- Estimate accuracy range indicates the degree to which the final cost outcome for a given project will vary from the estimate cost.
- Accuracy is expressed an a +/-% range around the point estimate after application of contingency with a stated level of confidence that the actual cost outcome would fall within this range (???)
- Contingency is established to address risks. The point estimate is the base estimate. (?) The point estimate plus the contingency (risk) will fall within a 60th %-ile confidence.
- Accuracy is dependent upon estimate deliverables and estimate maturity
- An estimate deliverable is the same as a project deliverable. (?) An example is a project summary. (?)
- Estimate maturity increases from estimate class 5 to class 1. A class 1 estimate is more accurate than a class 5. (?)
- Estimate accuracy is usually evaluated in conjunction with some form of risk analysis process

The gasoline analogy:

- My estimate range for regular gasoline in Olympia is between $2.89 and 2.99/gal based on research
- My point estimate is $2.94. (?) My variance is +/- 1.7%. That is not an accuracy estimate.
- However, I believe my accuracy is very high. "Accuracy is dependent on estimate deliverables." What is my estimate deliverable in this case? My research results?
- If I'm estimating my gas budget for the rest of the year, my accuracy goes down. I can determine from history how many miles I drive in a year. But I don't know if the Fed gas tax will increase or how much, if I'll be driving out of town, etc, etc. I need to add a risk contingency to cover unknowns based on risk analysis. I could estimate likelihood and impact for each factor and do a Monte Carlo simulation. How do I estimate my accuracy?
- So I figure I will drive another 7000 miles this year at $2.94/gal and 27 miles/gal ($762). I add $100 contingency for my known unknowns. I inflate this number to the mid-point of expenditure. Let's say, $900. My accuracy isn't +/- $100. It must have something to do with the accuracy of my assumptions: my miles driven assumption, my mileage assumption, inflation assumption, contingency assumptions. What's my estimate accuracy? How do I calculate it?

Any help is appreciated.

Figure 3.53   The Program Manager's Request

The estimate's range depends on the range of each item and how they depend on each other and their accuracy depends on the number of errors and their magnitude. For example, if we estimate a project with only two items it may result in three cases:

Case 1. The initial estimate looks like the data presented in the following table:

Item name	Low	High	Most likely	Range rate
		Million		
"A"	9	11	10	10.0%
"B"	16	24	20	20.0%
Total	25	35	30	16.7%

Case 2. New data is available and item B is estimated at $1 million instead of $20 million. The input range remains the same, as a percentage. The new total is given in the following table:

Item name	Low	High	Most likely	Range rate
		Million		
"A"	9	11	10	10.0%
"B"	0.8	1.2	1	20.0%
Total	9.8	12.2	11	10.9%

It looks like the total's range is narrower. Does it mean that the Case 2 estimate is more accurate? At this time it may be because the data is more recent but it may not be, since we do not know anything about its quality.

Case 3. New data keeps coming in and item B is estimated at $400 million. The new total is given in the following table:

Item name	Low	High	Most likely	Range rate
		Million		
"A"	9	11	10	10.0%
"B"	320	480	400	20.0%
Total	329	491	410	19.8%

The total cost range changes from one case to the other. But this doesn't say anything about the estimate's accuracy. It makes sense to assume that Case 3 is more accurate than the other two cases because it has the most recent data. But Case 3 has the largest range.

So association of the estimate's range and the estimate's accuracy doesn't work. As a matter of fact, it may be misleading.

In other words, an item defined by a wide range may be more accurate than the other item defined by a narrow range. Stated another way: The accuracy of an estimated item is not driven by the magnitude of the range associated with it. It is driven by the value of the data that defines the estimate. The same is valid when the total project cost is estimated. The RBE solicits data from internal and external SMEs to increase the quality of data that will be used.

- *Estimate accuracy range indicates the degree to which the final cost outcome for a given project will vary from the estimate cost. From AACE Recommended Practice No. 17R-97 p.3.* —We are not sure what this means. The term *estimate accuracy range* is new to us and we think that it deserves more explanation. It may be useful but at this time we are not aware how it should be evaluated.

- *Accuracy is expressed as a ±% range around the point estimate after application of contingency with a stated level of confidence that the actual cost outcome would fall within this range. (???)* —We are not sure what this means. We think that the issue is more complex than is implied by this statement. For example, the process of risk-based estimating (RBE) provides for the estimate a range and shape (a distribution). The accuracy of that estimate depends on the quality of the inputs and it has nothing to do with the estimate's range. As presented earlier, we would not recommend any association between the magnitudes of the range of an estimate and the measurement of its accuracy. *Also from 17R-97 p.3 (second sentence of same paragraph): The range is the accuracy range not the estimate range.*

- *Contingency is established to address risks. The point estimate is the base estimate. (?) The point estimate plus the contingency (risk) will fall within a 60th %-ile confidence.* —This may need better definition. In our opinion, once a risk-based estimate is employed, contingency loses its meaning and it should be replaced by the term *risk reserve*. AACE uses *contingency* so we stuck with its terminology even though we use it differently.

- *Accuracy is dependent upon estimate deliverables and estimate maturity.* —This statement may be true if "deliverable and maturity" are related to possible errors.

- *An estimate deliverable is the same as a project deliverable. (?) An example is a project summary. (?)* —I am not sure what this means. If an estimate deliverable is the same as a project deliverable why introduce this term (estimate deliverables)? Estimate deliverable is used in 17R-97 to quantify the level of estimate maturity [more later].

- *Estimate maturity increases from estimate class 5 to class 1. A class 1 estimate is more accurate than a class 5. (?)* —Okay. You are saying that once the project matures someone should expect better accuracy and perhaps a narrower range. Again, do not connect the estimate's accuracy to its range.

■ *Estimate accuracy is usually evaluated in conjunction with some form of risk analysis process.* —This may be wrong. As we have written before, an estimate's accuracy depends on the quality of data used to develop the estimate. Risk analysis increases the awareness of what may happen, how much it may cost, and how long it may take to finalize a project.

The gasoline analogy:

■ *My estimate range for regular gasoline in Olympia is between $2.89 and 2.99 per gallon based on research.* —Okay

■ *My point estimate is $2.94. (?) My variance is ± 1.7 percent. That is not an accurate estimate.* —Okay. At this point you have a base cost and the variability associated with it. "Variability" shall have a neutral meaning. No events that may change its values are captured in the range calculated.

■ *However, I believe my accuracy is very high. "Accuracy is dependent on estimate deliverables." What is my estimate deliverable in this case? My research results?* —Yes. Your estimate accuracy is very high. The number $2.94 represents the base estimate at that point in time. Tomorrow it may be different. The 1.7 percent is the variability in the base for the unit price. Even if you would like to buy some gas today you do not know precisely how many gallons you need, which may increase the variability for today's cost of the gasoline to fill up your tank. We can discuss this subject for hours . . .

■ *If I'm estimating my gas budget for the rest of the year, my accuracy goes down.* —Okay. *I can determine from history how many miles I drive in a year. But I don't know if the federal gas tax will increase or by how much. I'm not sure if I'll be driving out of town, etc., etc. I need to add a risk contingency to cover unknowns based on risk analysis. I could estimate likelihood and impact for each factor and do a Monte Carlo simulation. How do I estimate my accuracy?* —The accuracy will depend on how reliable your history is, how reliable the forecast of inflation of federal gas tax is, and on how reliable your assumptions are such as "I'm not sure if I'll be driving out of town, etc., etc." Running a risk analysis and determining a range and shape of the estimated cost will not increase the estimate accuracy. The estimate's accuracy will increase only if you manage to get better information such as: make sure your driving history is accurate, investigate the forecast of federal gas tax, clarify your trips, and be specific on the "etc., etc."

■ *So I figure I will drive another 7000 miles this year at $2.94/gal and 27 miles/gal ($762).* —Okay. *I add $100 contingency for my known unknowns.* —Okay. This represents an "Allowance" to cover the cost of items you know you need but don't know how much they will cost. *I inflate this number to the mid-point of expenditure, let's say, $900. My accuracy isn't ± $100.* —Okay. The estimate's accuracy has nothing to do with $100. *It must have something to do with the accuracy of my assumptions: my miles driven assumption, my mileage assumption, inflation assumption, contingency assumptions . . .* Contingency assumptions are not the right term for known unknowns, it is preferable to call them allowances and you need to add risks. *What's my estimate accuracy? How* *do I calculate it?* **Great questions! If a Monte Carlo simulation is run then the estimate will have a range and shape. The accuracy of this range and shape depends on the quality of inputs. The quality of some inputs is possible to measure and the quality of other inputs will be difficult to measure. It may be a formula that will give a sense of the result's accuracy, but this may be the theme of few doctoral research papers.**

The dialog ends here.

As you notice, the PM had the answer: "It must have something to do with the accuracy of my assumptions: my miles driven assumption, my mileage assumption, inflation assumption, contingency assumptions." This is it. The quality of data used in the estimate dictates the estimate's accuracy.

CONCLUSIONS

The risk-based estimate is a valuable estimating process that may assist the project manager with risk management and project cost and schedule estimating. The RBE gives management a sharper and far more realistic long-distance view of the prospects awaiting their projects. The data that RBE provides to management allow a reasonable understanding of the project boundaries related to its scope, delivery date, and how much it might cost.

Through its candidates for mitigation and project risk map, the RBE provides excellent data for developing a sound project risk management plan. Once a risk management plan is developed the RBE may provide the effects of implementing the plan. The process could cycle again and again as many times is deemed necessary.

When RBE is done correctly it minimizes the number of surprises and, most importantly, helps the project manager to reduce the threats impact and maximize the opportunities effect by providing quantified data of risk events. In other words, it increases chances of successfully delivering the project.

RBE creates opportunities to study what-if scenarios using a rigorous and statistical approach. The individual risk or group of risks must be studied and the impacts to the project must be understood. Furthermore, the what-if scenario may be employed for analyzing different project alternatives which may greatly help decision makers on selecting a preferred alternative.

RBE allows reasonable control over the project's estimate through project risk management. It provides advanced warning to the project manager so he or she may make proactive and informed decisions about dealing with project uncertainties.

A huge benefit of the RBE process consists in improving the project communication and transfer of information among the project team members, stakeholders, and other entities. This real and important benefit, which may be underestimated, was presented at the beginning of this chapter: Once people go through a good RBE exercise they appreciate the real soft value of it.

Realistic contingency planning (risk reserve) is made possible since it considers the effect of positive and negative events that may affect the project. This is a matter of understanding the project and the level of risk tolerance that the organizations have. Perhaps a new chapter may be written about it but at this time the book doesn't include it.

The chapter presents two approaches to RBE—Keep It Simple, Smarty and professional sophistication—and it recommends eliminating the second one, no matter how attractive it is. When professional sophistication runs high, the possibility of having problems with the analysis is also high. The most detrimental effects to the RBE process are generated by hidden components of professional sophistication such as: (1) a large number of variables (risks), (2) poor risks conditionality (dependency and/or correlation), and (3) vague definition of a risk's distribution. While KISS may guard the analysis against some of these fallacies, the KISS principle is not a panacea for risk analysis. The role of the risk analyst and/or risk elicitor is crucial on any risk assessment and risk analysis.

Base uncertainty recognizes that at any point in time nobody can estimate for sure how much a project is actually going to cost two or five years from now, even if the project is delivered without the occurrence of project-specific events (i.e., risks). We may only forecast a range of probable cost that is based on the information available at the time of the estimate.

The probability bound, introduced in this chapter, is an excellent concept that captures both components of the base uncertainty: (1) epistemic and (2) random. Each component is crucial in defining reliable and robust base uncertainty. It is up to knowledgeable professionals to define these two components for each individual project. The attention, or lack thereof, given to base uncertainty may decide if the analysis will help guide project decision making or hinder it in terms of meeting the project's goals.

It is recommended that RBE uses simple distributions that may be easily understood by SMEs, project managers, and stakeholders. The discrete distribution may be used on rare occasions and the Pert and triangular distributions should be dominant if not exclusive.

The minimum and maximum values are the best choice to be used when the distribution range is defined. There are situations when an SME may be reluctant to give a value for the minimum and maximum and in this case we recommend using the LOW and HIGH terms and identify their meaning. In both cases it is important that the SMEs understand exactly the meaning and significance of what they estimate. To the extent possible, it is recommended that the distribution shape be displayed in front of the SMEs so that they will understand exactly what they are estimating.

The RBE recognizes the indivisible value of the project's triad and the necessity of treating the cost and schedule together once the scope is defined. The integrated approach to project cost and schedule risk analysis leads to better results because it binds together cost and schedule for every single situation and it provides for robust and comprehensive understanding of the project's prospects.

The RBE is one of the most versatile methods of estimating and risk assessment. It can be applied to projects from small values to mega values; it can be applied to projects from the planning stage to the construction stage; it can be applied to projects from the simplest one to the most controversial one. It is just a matter of understanding the project and the commensurate level of efforts to achieve the goal of the action.

ENDNOTES

1. S. Lichtenberg, *Proactive Management of Uncertainty—Using the Successive Principle*, www.lichtenberg.org/, and B. Flyvbjerg, M. Holm, and S. Buhl, "Understanding Costs in Public Works Projects," *APA Journal*, 68(3) (Summer 2002), pp. 279–295.

2. K. Humphreys, "Risk Analysis and Contingency Determination Using Range Estimating," AACE International Recommended Practice No. 41R-08 (2008); .S. Lichtenberg, Oct. 2005, "How to Avoid Overruns and Delays Successfully—Nine Basic Rules and an Associated Operable Procedure," *ICEC Internet Journal*, 18(4), www.icoste.org/Roundup0406/Lichtenberg.pdf.

3. O. Cretu, "Cost and Schedule Estimate—Risk Analysis for Transportation Infrastructures," International Mechanical Engineering Congress and Exposition, November 13–19, 2009, Lake Buena Vista, FL; WSDOT, Oct. 2008, "Guidelines for CRA-CEVP Workshops"; www.wsdot.wa.gov/Projects/ProjectMgmt/RiskAssessment/; WSDOT, "Cost Estimating Guidance for WSDOT Projects," M 3034.0 (November 2008); www.wsdot.wa.gov/publications/manuals/fulltext/M3034/EstimatingGuidelines.pdf; O. Cretu and T. Berends, "Risk Based Estimate Self-Modeling" 2009 AACE International TRANSACTIONS, Seattle, WA, June 28–July 1, 2009; Risk-Based Self-Modeling Spreadsheet, www.cretugroup.com; O. Cretu, T. Berends, R. Stewart, "Reflections about Base Cost Uncertainty" Society for Risk Analysis Annual Meeting 2009, Risk Analysis: The Evolution of a Science, December 6–9, 2009, Baltimore, MD; J. Reilly, M. McBride, D. Sangrey, D. MacDonald, and J. Brown, "The Development of a New Cost-Risk Estimating Process for Transportation Infrastructure Projects," *Civil Engineering Practice*, 19(1); O. Cretu, "Risk Based Estimate of Transportation Infrastructures," The International PIARC Seminar—Managing Operational Risks on Roads November 5–7, 2009, Iasi, Romania; K. Molenaar, S. Anderson, C. Schexnayder, "Guide on Risk Analysis Tools and Management Practices to Control Transportation Project Cost," *NCHRP Report 658*, Transportation Research Board, 2010; W. Roberts, T. McGrath, K. Molenaar, J. Bryant, "Guide for the Process of Managing Risk on Rapid Renewal Projects," Strategic Highway Research Program, SHRP 2, Project 09, 2010.

4. O. Cretu, T. Berends, R. Stewart, V. Cretu, *Risk-Based Estimate—Keep It Simple*, International Mechanical Engineering Congress and Exposition, November 13–19, 2010, Vancouver, BC.

5. B.J. Garrick et al., *Quantifying and Controlling Catastrophic Risks*, Amsterdam: Elsevier, 2008; E.T. Jaynes, *Probability Theory; The Logic of Science*. Cambridge, U.K.: Cambridge University Press, 2003; S. Plous, *The Psychology of Judgment and Decision Making*. New York: McGraw-Hill, 1993; R. Nofsinger, *Investment Madness: How Psychology Affects Your Investing . . . and What to Do About It*. Upper Saddle River, NJ: Pearson Education, 2001; R.B. Cumming, "Is Risk Assessment a Science?" *Risk Analysis*, 1(1):1–3 (1981); D.A. Amir , *Statistics: Concepts and Applications*. Homewood, IL: Richard D. Irwin Inc., 1995; P.L. Bernstein, *Against the Gods: The Remarkable Story of Risk*. New York: John Wiley and Sons, 1996.

6. Transportation Estimators Association, (TEA) www.tea.cloverleaf.net/.

7. D. Karanki, H. Kushwaha, A. Verma, and S. Ajit, "Uncertainty Analysis Based on Probability Bounds (P-Box) Approach in Probabilistic Safety Assessment," *Risk Analysis*, Vol. 29, No. 5, 2009.

8. Ibid.

9. S. Ferson and J.G. Hajago, "Arithmetic with Uncertain Numbers: Rigorous and (often) Best Possible Answers." *Reliability Engineering and System Safety*, Vol. 85 July–Sept. 2004, pp. 135–152.

10. T.W. Tucker, S. Ferson, "Probability Bounds Analysis in Environmental Risk Assessment." *Applied Biomathematics*, 13–17, 2003.

11. M. Bruns, and C.J.J. Paredis, "Numerical Methods for Propagating Imprecise Uncertainty." Proceedings of IDETC 2006, September 10–13, Philadelphia, 2006.

12. H.C. Frey and R. Bharvirkar, "Quantification of Variability and Uncertainty: A Case Study of Power Plant Hazardous Air Pollution Emissions." In *Human and Ecological Risk Analysis*, Dennis J. Paustenbach, ed., pp. 587-617. Hoboken, NJ: John Wiley and Sons, 2002.

13. D. Vose, *Risk Analysis—A Qualitative Guide*. Hoboken, NJ: John Wiley and Sons, 2009.

14. Carveth Read, *Logic—Deductive and Inductive*. London: Grant Richards, 1898.

15. L.S. Savage, *The Flaw of Averages—Why We Underestimate Risk in the Face of Uncertainty*. Hoboken, NJ: John Wiley and Sons, 2009.

CHAPTER 4

RISK ELICITATION

IMPORTANCE OF ELICITATION FOR PROJECT RISK MANAGEMENT

The previous chapter describes RBE with its two main components: (1) base estimate (cost and schedule) and (2) risks. The base estimate is produced by the project team and the base cost and schedule team reviews and validates it. Risks are identified and quantified during the process of RBE through a collaborative effort conducted by the risk lead. The risk lead elicits from internal and external subject matter experts (SME) information about what may happen during the life of the project that may significantly change the project cost and schedule. The elicitation process is similar to "normal" meeting facilitation but it is more challenging because the elicitation must overcome a series of biases that the participants may have. Later in the chapter we will describe the most damaging biases that may affect the RBE results.

The elicitation of risk information is a demanding process that produces the project's risk register which provides all of the risk data concerning the project and is then used by the risk modeler to develop his or her model. Risk elicitation is the part of the cost estimating process where the distinction is made between a deterministic estimate and a risk-based estimate. Someone may say: "If I associate a deterministic estimate value to a distribution (range and shape) then I can obtain results that will be similar to an RBE, but with less time and effort." The last part of this statement is true but the first part is in error. The statement's fallacy lies within the unknown that is hidden in the tails of the distribution.

In rare cases, when the distribution associated to the deterministic estimate is the correct one, having the distribution alone does not help the decision makers. The decision process is based on the information about what is included in the tails' distribution. The RBE provides these kinds of data, which are collected through the process of risk elicitation.

There are three main ways of conducting risk elicitation:

1. One-on-one interviews
2. Large group
3. Small group

One-on-One Interview

Conducting individual interviews of an SME seems to be the simplest way of eliciting risks whereby the risk elicitor tries to have a direct discussion with a knowledgeable person. The elicitation is productive when the interviewee is outspoken, but there are cases when an individual doesn't feel comfortable about making a decision on risk's probability of occurrence and its impact. There are other situations when the data provided by one interviewee conflicts with the data about the same risk provided by another, equally qualified interviewee. In many cases the interviewee may find excuses and defers a direct discussion. As a result of these behaviors, the one-on-one risks interview lacks group synergy, which is often an important factor of risk elicitation.

Large Group Elicitation

Large workshops with many diversified specialty groups can create "too much group synergy." Too many times valuable workshop time is wasted by trivial specialized discussions generated by persons without knowledge or who are not experts in a particular subject. When workshops have large and diversified groups of people, the group synergy can quickly develop into "synergetic chaos." Under these conditions, risk elicitation becomes unproductive and often delivers ill-defined risks.

The risk elicitor's job becomes a challenge when the group is larger than 20 people. Worse still, when the number of people in attendance is more than 30, the entire process of risk elicitation can become an exercise in futility. The authors have noticed that people tend to be inclined to discuss minor issues that contribute little value to the process and that this tendency is amplified and more difficult to control when the number of workshop participants is higher than 20. In general, people are afraid of identifying the extreme events, because of the fear of looking dumb. These extreme events are the ones that need to be discussed because they are the ones that will have a significant impact on the outcome of the project.

This is not to say that large group elicitation cannot be effective, only that it requires a risk elicitor with strong facilitation skills and who can also command control of the room.

Small Group Elicitation

Small group (includes risk elicitor, project team, and SMEs) discussions have proven to be the most productive procedure for the elicitation of risks. Small group discussions better accommodate candid, and sometimes controversial, debate among experts without the worry of misunderstanding. The authors commonly refer to the small group format as an "advance risk elicitation interview." Each advance risk elicitation interview is focused on one or two specific areas of interest (geo-tech, structures, environmental issues, design issues, and so forth). During these small group discussions, the majority of risks are identified and quantified.

The advance risk elicitation interview approach presents the advantage of having the right people involved to debate and assess the project risks. It has been observed that having small groups of SMEs focused on project-specific areas creates better synergy than large and diversified groups. Usually the advanced risk elicitation interviews take place prior to the workshop. Under these conditions, the SMEs should have time to reevaluate their probabilities; recalculate the impact values; and so on. The advanced risk elicitation interview contributes to increasing the acceptance and credibility of the cost risk analysis from within.

The credibility of the RBE is essential to the process as a whole and to the quality of what the process provides. The advanced risk elicitation interview nourishes the RBE credibility from within because it allows participants to digest the information they provide for the analysis. When risk elicitation is performed only during the workshop, in front of 20 to 30 strangers, an SME is more prone to feel insecure and uncomfortable with providing probabilities and impacts (low, high, and most likely) for risks that they just learned about. SMEs must be afforded the time to thoroughly consider risks rather than have to "shoot from the hip." Their accuracy is almost always better when they are given time to take aim. Furthermore, the advanced risks elicitation interview helps the full workshop discussion to progress smoothly since the experts have had sufficient time to make up their minds, and agreements on risk assessments are completed in a timely fashion.

Regardless of how risks are elicited, a good risk elicitor must have strong people skills to be able to communicate and paraphrase what SMEs may bring up. The risk elicitor is the driving person during the elicitation process (one-on-one, advance risk elicitation interviews, or workshop) and at the same time needs to stay neutral. He or she must avoid, at all cost, the perception of "force feeding" while guarding against people's biases.

The risk elicitor must understand the project risk environment both in a holistic form and a detailed form. It is important that they have a general working knowledge of the project that they are involved with as they must be able to understand what is being said and have some concept of the significance of the information being elicited. The risk elicitor should use only plain language when guiding the SMEs through the tasks and when translating the SMEs' inputs, so that the data is ready for inclusion in the model.

ELICITATION AND BIASES

It is essential to develop an understanding of how the concepts of *elicitation* and *cognitive bias* influence decision making within the context of the analysis and quantification of risk. These concepts are also relevant to any type of group communication or decision-making process. Let us first examine the definitions of these two key concepts.

Elicitation is essentially the process whereby one draws out information from another. This process usually occurs through an iterative process of questions and answers. What is important to understand here is that the way in which questions are posed, or "framed," directly influences the nature of the answer or response.

Cognitive bias describes a distortion in the way we see reality. Cognitive bias can be thought of as a filter that alters the way in which we interpret our environment. There are a multitude of cognitive biases that can alter how we perceive the world around us and, in turn, affect how we make decisions based on the misinterpretation of information.

Cognitive bias refers to any of a wide range of observer effects identified in cognitive science and social psychology. They include very basic statistical, social attribution, and memory errors that are common to all humans. Biases can degrade the reliability of our observations and memories. Social biases, usually called attribution biases, affect our everyday social interactions, while biases related to probability and decision making can significantly affect the very tools and techniques that have been designed to minimize such biases. Well over a hundred specific cognitive biases are known to exist. Imagine how many more must exist that have not yet been formally identified and studied.

Cognitive biases can play a significant role in the elicitation of risk, especially as this relates to our ability to define ranges and identify subjective probabilities. A great deal of research has been conducted over the years on this topic, and there appear to be two major conclusions that can be drawn from it: that the human mind has only limited information-processing capacity and that the nature of a task has a great impact on the strategies that are chosen to deal with the task. The consequences of these phenomena are:

- Our perception of information is not comprehensive but selective. Since we are only capable of apprehending a small part of our environment, our anticipations of what we will perceive determine to a large extent what we actually perceive.
- As we do not have the capacity to make what one might call "optimal" calculations, we make much use of heuristics and cognitive simplification mechanisms.
- Since we cannot simultaneously integrate a great deal of information, we are forced to process information in a sequential fashion.[1]

Each of these consequences exposes us to the effect of cognitive biases to varying degrees. Knowledge of these biases is essential so that the risk elicitor can pose questions in a manner that will minimize their effect to the maximum extent possible.

Heuristics

There are a number of cognitive biases worth discussing that play the greatest role during the risk elicitation process. These biases center on the psychological and physiological "blind spots" of the individual participants including the risk elicitor. Some are based on the rules of thumb, or heuristics, that simplify the decision-making process and enable us to make quick choices in our daily lives. While these heuristics are often very helpful in dealing with an ever more complicated world, they can also create mental roadblocks that prevent objective decisions. Three of these heuristics include:

- Anchoring heuristic
- Availability heuristic
- Representativeness heuristic

The famous behavioral psychologists Amos Tversky and Daniel Kahneman explored and popularized these three heuristics in a 1974 article featured in *Science* magazine.[2]

Anchoring Heuristic

In this section we are going to explore the anchoring heuristic. But before we begin, you will first need to participate in a brief exercise. Let us assume that you are in a room with three other people—Bob, Sarah, and Jamal. I am going to ask the four of you a question. Assume that the other three provide their answers before you do. Here is the question:

Suppose you randomly pick one of the countries represented in the United Nations. What is the probability that it will be an African nation?[3]

Kahneman and Tversky posed this exact same question to groups of individuals in a controlled environment. Before they made their estimate, they were given a random anchor that was generated by a spinning wheel that contained the numbers 0 through 100. The wheel was rigged so that half of the participants received 10 for their anchor and the other half received 65 for their anchor. They found that when 65 was the anchor, the mean result was 45. When 10 was the anchor, the mean result was 25. They identified this phenomenon as the anchoring heuristic.

The anchoring heuristic describes the tendency for people to explain or describe an event by fixating on the first number or evidence that they hear. After forming an initial belief, people tend to be biased against abandoning it. Referring back to the previous example, when people saw 65, they tended to anchor to that number and then adjust up or down accordingly. Adjustments based on an anchor can be inadequate if the anchor deviates significantly from reality. This suggests that you can bias people's estimates if you provide the initial anchor.

As we will explore later in this chapter, people seem to be ill-equipped when it comes to estimating the extremes of distributions. It is likely that the anchoring heuristic plays a role in this and further exacerbates this phenomenon. One way to counteract this is when eliciting the ranges of impacts: first identify the *maximum* or *worst case* impact, therefore setting the initial anchor at a higher position whereby adjustments will be made down from this anchor to identify the most likely impact value.

The anchoring heuristic can be minimized by utilizing group evaluation techniques rather than relying on individual evaluation methods. Drawing on the experiences and perspectives of people representing different disciplines and philosophies will help to expand the discussion and keep this heuristic from eclipsing ideas that deserve further consideration.

The following strategies should be considered to minimize the influence of the anchoring heuristic during risk elicitation:

- Explain the basic mechanics of the anchoring heuristic to the group prior to eliciting risks.
- When possible, state the "status quo" position when asking a question.
- Try to frame questions based on facts rather than assumptions.
- Identify the maximum or worst-case impact first.

Availability Heuristic

In this section we will discuss the availability heuristic. Choose from the following pairs after reading the following question.

What are you most likely to die from over the course of your life if you live in the United States?

- Earthquake or bee sting?
- Accidental suffocation or drowning?

The majority of people select earthquakes and drowning over bee stings and suffocation. This is puzzling, especially when the probability of dying from the second set of misfortunes is, in most cases, twice as likely. The statistics were compiled by the U.S. National Safety Council and identify the probabilities of dying from the various calamities identified above (see Table 4.1).

This phenomenon is known as the availability heuristic. It describes the influence that cognitive visualization has on critical thinking. The more vivid an image is within our mind, the stronger the influence it has on our critical thinking. As a result, when we are faced with choices, we will tend to be biased toward the more vivid image. Murders are more vivid than suicides, and are certainly more prolific from the standpoint of the media and entertainment industries. There are far more images of graphic murders in our heads than suicides, therefore we may be predisposed to think that murders are more frequent than suicides. Similarly, earthquakes can result in wide-ranging and catastrophic damage. Bee stings, by comparison, seem rather innocuous although you are twice as likely to die from a bee sting as an earthquake.

An important corollary finding to the availability heuristic is that people asked to imagine an outcome immediately perceive it as more likely than those who were not. And that which was vividly described is viewed as more likely than that which was provided a much duller description.

TABLE 4.1   Lifetime Statistical Probabilities of Causes of Death for U.S. Residents

Cause of Death	Odds	Cause of Death	Odds
Bee Sting	1 in 46,477	Earthquake	1 in 103,004
Suffocation	1 in 646	Drowning	1 in 1,064

Safety is often an important performance attribute for many types of projects. It is also probably the most obvious trigger for the availability heuristic because a project or facility that is perceived as "unsafe" conjures up vivid images of the consequences.

Often people have great difficulty in conceiving risk impacts without a visual anchor. People are visual thinkers, and a skilled risk elicitor can utilize this heuristic to his or her advantage by helping others imagine a risk event unfolding. This is especially effective when trying to conjure up a worst-case scenario.

For example, one might explore a worst-case scenario by describing the risks associated with damaging a water main during excavation. Imagine that while excavating a new utility trench for a roadway project, the backhoe ruptures a water main that was supposed to be located 10 feet away. Before the water can be shut off, the rupture creates a sinkhole that blocks off the road and swallows a Mercedes Benz. The project is delayed for a week while emergency repairs are made. (Let's assume that the contractor is insured, so the monetary impacts are absorbed.)

Providing a context-rich description of a risk event makes it more available to the mind's eye, and improves our ability to use our creativity to articulate and quantify impacts.

The following strategies should be considered to minimize the influence of the availability heuristic during group elicitation:

- Explain the basic mechanics of the availability heuristic to the group prior to eliciting responses. Do this by asking questions similar to the ones in the previous example.

- Try to frame risks that will appeal to logic rather than emotion.

- Be especially sensitive to emotionally charged risks, especially those that deal with safety or life and death consequences.

- Explain how visualizations, especially those related to recent events, can influence our perception of probability.

- Use the availability heuristic in a proactive way to stimulate visualization in identifying risk impacts.

Representativeness Heuristic

In this section we will explore the representativeness heuristic. We will begin with a simple exercise to help demonstrate this heuristic in action. Let me describe to you a man named Jack. Jack is 45 years old, married, and has four children. He is generally conservative, careful, and quiet. He shows no interest in political and social issues and spends most of his free time building models with his sons, reading, and solving mathematical puzzles. Based on the information provided, is Jack a lawyer or a salesman?

Imagine further that Jack is attending a conference with 99 other men. Of the total number of 100 men present, 30 are engineers and 70 are salesmen. Based on what you know about Jack, what would you estimate to be the probability that Jack is one of the engineers?

If we relied on the data in the previous paragraph, the answer would be 30 percent. Did you select a different number? If you did, your answer is at least partly based on the description of

Jack. This is an example of the representativeness heuristic. One of the key biases emanating from this heuristic is the tendency to overlook base rates in favor of case-specific information. This means that information unique to the situation is likely to have far greater influence than historical or parametric data that is potentially misleading.

The use of this heuristic can, however, systematically lead one to make poor judgments in some circumstances. Other examples include:

- The belief in runs of good and bad luck in games of chance. This particular incarnation is also known as the *gambler's fallacy*.
- People will often assume that a random sequence in a lottery (12, 19, 57, 23, 8, 31) is more likely than an arithmetic sequence of numbers (5, 6, 7, 8, 9, 10).
- If two salespeople from a large company both displayed aggressive behavior, the assumption may be that the company has established a policy of aggressive selling, and that most other salespeople from that firm will also engage in aggressive techniques.

In summary, people tend to estimate the probability of an event by how similar the event is to the population of events it came from and whether the events seem to be similar to the process that produced them.

The following strategies should be considered to minimize the influence of the representativeness heuristic during group elicitation:

- Explain the basic mechanics of the representativeness heuristic to the group prior to eliciting responses.
- Recognize that this bias affects our ability to perceive the probability of events.
- Raise awareness of this heuristic within the group by stating an absurd stereotype that is relevant to the group. For instance, if the group were composed of architects, you could state, "Architects don't care about what buildings cost—they only care about what they look like." Be mindful, though, to avoid offensive references!

The risk lead should seek to develop an understanding of these heuristics and biases and try to recognize them when they occur, as they can quickly, and unfairly, derail ideas that may otherwise prove to have merit. The representativeness heuristic is often the worst, and most unfair, of the three heuristics discussed in this chapter. In its most destructive form, this heuristic is really nothing more than prejudice based on broad stereotypes. The risk lead must use tact in disarming this behavior.

Overconfidence Effect

The overconfidence effect describes the tendency for individuals to place much greater confidence in the reliability of their judgments or estimates than they should. It has been shown through numerous studies that people are not very good at estimating the ranges of unknown quantities. This is an important phenomenon to consider when eliciting ranges of impacts for risk events.

In one important study, researchers asked Harvard Business School students to estimate unknown quantities such as the percentage of their classmates who preferred bourbon to scotch and the total egg production in the United States for a given year. Naturally, the students were uncertain about these facts and did not have access to this information at the time of the study. The researchers also asked them to specify the lower (1 percent) and upper (99 percent) bounds of the estimate such that they were 98 percent certain that the true value was somewhere between these two extremes. If students had specified intervals that were sufficiently wide given their uncertainty, then 98 percent of them should have captured the true value and only 2 percent should have failed. The results proved that 42 percent failed to capture the true value—which is 40 percent higher than what should have been expected. If the true estimate lies outside the specified interval too often, that's evidence of overconfidence. If it lies inside too often, that's evidence of underconfidence. The result of this study indicates that the students offered intervals that were too narrow, an indication of overconfidence.[4]

Experts are not immune to the overconfidence effect. In one study, seven very experienced geotechnical engineers were asked to predict the height of an embankment that would cause a clay foundation to fail. Specifically, they were asked to identify the confidence bounds around their estimate of this failure that were wide enough to have a 50 percent chance of enclosing the true height. Interestingly, none of the bounds specified captured the true failure height.[5]

Another study reported physician estimates for the probability of pneumonia for 1,531 patients examined that displayed a similar symptom (in this case, a cough). Eighty-eight percent of the diagnoses made by the physicians studied indicated that the patients had pneumonia when in fact the actual number was less than 20 percent.[6]

Both of these examples provide fairly striking evidence of how overconfidence can influence estimates of subjective probabilities. Despite the poor results of these studies, it should be noted that experts do perform better than nonexperts in assessing ranges and probabilities, so long as they are dealing with areas in which they are indeed expert.[7]

The key during risk elicitation with respect to the overconfidence effect is to stress the importance of establishing a sufficiently wide range for the lower and upper bounds (i.e., minimum and maximum) when establishing impacts for risk events. It is very common for participants involved in a risk elicitation session to identify a range that is too narrow. The risk elicitor should be aware of this and seek to draw the group's attention to this when this occurs. Additional discussion should ensue to stimulate creative thought, especially with respect to worst case scenarios. Impact visualization is an excellent exercise to engage in to counteract this tendency as discussed previously in this chapter.

Motivational Bias

Motivational bias occurs when people have an incentive to reach a predetermined conclusion or see things in a specific way. This bias is one of the more dangerous ones that can affect the elicitation of subjective probabilities and impacts from experts. There are numerous reasons for why the motivational bias can occur. Examples include:

- A person may want to influence a decision toward a specific outcome and therefore adjust up or down the probabilities and impacts of a risk event.
- A person may perceive that he or she will be evaluated based on the outcome of a decision and therefore might tend to be conservative in his or her estimates.
- The person may want to adjust up or down the degree of uncertainty that he or she actually believes is present in order to appear knowledgeable or authoritative. Similarly, the adjustments could also be made to avoid the appearance of being ignorant or powerless.
- The expert has taken a strong stand in the past and does not want to appear to contradict him- or herself by producing a distribution that supports contrary positions.

Experts are often the best resource for eliciting judgments on uncertain events; however, the risk elicitor should be aware of the influence of self-interest. Experts don't generally become "experts" without having also experienced some tangible effect on their egos. The authors have seen this bias at play on many risk assessments. The risk elicitor must seek to frame questions in a way that relate to the facts, rather than in a way that will elicit responses that are likely to direct the focus on the self-importance of the expert.

A common issue that often arises during risk elicitation sessions is when the participants include both superiors and subordinates. Many times subordinates are afraid to speak their minds out of fear of contradicting or embarrassing their bosses, which could negatively affect their position or future prospects within the organization. Similarly, the authors have witnessed numerous instances where expert consultants are unwilling to challenge the assumptions of their clients for fear of jeopardizing current or future work.

In these situations, a strategy that can be effective is to first elicit the risks, probabilities, and impacts from subordinate or consultant groups independently without the presence of the superiors. The risk elicitor can then "sanitize" this information by making it anonymous before sharing it with the superiors. This approach can be very effective at opening the eyes of upper management to risks that they were unaware of while extracting higher quality information from the people who are more intimately aware of the issues.

Optimism Bias

The optimism bias describes the tendency of people to overestimate the probability of good outcomes and underestimate the probability of bad outcomes. This is a personal bias that colors the way we consider risks.

One excellent example of how an entire business sector has exploited this bias is the consumer credit industry. Credit card companies have exploited the optimism bias for decades by luring borrowers to accept seemingly favorable credit terms that usually possess a nasty trap. The trap is that if you make a late payment, the interest rates instantly jump from 7 to 23 percent. Or, in another common scenario, interest is effectively offered at 0 percent so long as the borrower pays off the principal amount by a certain date. If they do not, they are responsible for retroactively paying the accrued interest at a very high rate. These strategies are extremely effective because people believe that they will never miss a payment—in other words, they are overly optimistic.

This bias is important to consider when eliciting risk information from people who have a stake in the project. People tend to be overly optimistic about decisions in which they play an active role because they believe they can somehow better control uncertainty. A project manager is more likely to believe that a project under her control will be delivered on time and on budget than a similar project being managed by a coworker when, in fact, they share the exact same risk factors.

The Elicitation of Risk Impact and Probability of Occurrence

This chapter has presented three formats for the elicitation or risks as well as the cognitive biases that can affect them. Provided below is a discussion of the techniques and considerations for the elicitation of risk impacts and probabilities.

It is worth emphasizing the importance of giving the identification of opportunities a commensurate level of attention as the identification of threats. It is the experience of the authors that the majority of SMEs are more inclined to identify threats than opportunities. The risk elicitor should be aware of this and seek to stimulate discussion with respect to the identification of opportunities. One excellent means of stimulating this discussion is through the application of value engineering (VE). (The use of VE will be expanded on later in this chapter.)

Elicitation of Risk Impact

Once a reasonable understanding of the nature of the risk event is developed, the next step is to begin thinking about what might happen to the project should it occur. Usually there are a range of possibilities. For instance, if the project has a risk of potentially rupturing a water line during excavation, the impacts could vary depending on the location of the rupture, the time of day, and the extent of the damage to the pipeline.

What is recommended is that the team begins with the worst-case scenario. Stop for a minute and imagine the worst thing that could possibly happen. In the case of the ruptured water line, let's imagine that the ruptured pipe ends up washing out the subgrade beneath an adjacent street during rush hour, creating a small sinkhole in the roadway that also triggers a traffic accident. Possible? Yes. Likely? No. However, at this stage we are not concerned with probabilities, only impacts.

Teams need to stretch their minds a little here. Worst-case scenarios generally are pretty rare, but they can often be particularly devastating. Keep the team solidly focused on impacts. Sometimes, this can be a bit like a game so have a little fun with this.

Now that the team has visualized the disaster, think about what this might do to the project in terms of scope, schedule, and cost. What would it cost to repair the pipeline, roadway, and cover the property damage? How would it affect the construction schedule? Try to come up with an estimate for both cost and time. Often, the best we can do is to make an educated guess; however, by visualizing a real incident we can at least come up with an approximation. Another thing to consider here is, who will pay for the risk? Many times, insurance policies and construction bonds will absorb financial losses. Make a note of this, as such costs may be real but may not directly affect the project budget. Time losses, on the other hand, probably will affect the project in terms of time-related overhead and project escalation.

Once the team has imagined the worst-case scenario, switch gears and think about what the best case scenario is. That is, if the risk event happened, what is the smallest impact that could occur? Do not say "nothing!" It has to be something! What is it? Continuing with the pipeline example, let's say the best case scenario is that the pipeline is bumped by a backhoe and develops a small leak. Let's assume the existing pipeline can be inspected and repaired with a one-day delay. This seems reasonable—*something* happened, it just wasn't a big impact.

Once the team has identified the worst- and best-case scenarios, the team should now think about the *most likely* scenario. This should fall somewhere in between the worst and best cases. Going back to the pipeline risk, it seems likely that if a backhoe hit the pipeline, it would probably result in a break that would have to be replaced. The break would probably not create a sinkhole and would probably be quickly contained. In this case, the water main would have to be turned off, the damaged section of pipe repaired, and construction would resume. The ensuing delay might take a few days to resolve. Record the best, worst, and most likely scenarios in the risk register.

Elicitation of Risk Probabilities

Once we have established the potential impacts of the risk event, we are now in a proper frame of mind to consider probability. This is the trickiest aspect of risk analysis and the one that gives people the most trouble. The thing to keep in mind here is that risk, by definition, is uncertainty. We don't know, nor can we know, what the "real" probability of a risk occurring is. That is the nature of risk—if we knew, then it wouldn't really be a risk.

It sometimes surprises me how often I have to remind people of this. Technical folks (i.e., engineers) can really have a hard time grasping this because they deal in the world of facts. Probability shifts us into the realm of the unknown and that gets us out of our comfort zone. This is especially true when people are asked to assign an actual number to a probability (i.e., 20 percent, 75 percent, whatever) because then it provides the illusion that an educated guess is now a matter of fact. It is not. However, in the face of uncertainty, an educated guess is infinitely better than no guess at all.

We do have some tools at our disposal to help us with this. Sadly, it isn't a magic crystal ball. Ironically, a popular brand of risk analysis software is named just that, Crystal Ball, which deals with quantitative risk analysis. The tools we have for risk elicitation include:

- **Personal experience**—Fortunately, we have a large, multidiscipline risk management team from which we can draw upon their collective experience. Much of our predictions with respect to both impact and probability will be derived from such experience. The past is usually a good indicator of the future in terms of projections.

- **Lessons learned**—We can draw upon the experience of past projects that are similar in nature. This represents experience that may lie outside of the team's own experience, but is available either anecdotally through others or through documentation. Use this information if you can get it!

- **Expert opinion**—Another tool we have is the expert opinions of others. If the team does not have the experience it feels is necessary, then members should seek out others who might know better. Once the information is received, have the team review it. Does it make sense? Are there any faulty assumptions? Never take anything as gospel and consider all information carefully.

- **Empirical data**—There is a great deal of data available that can show historical trends. This is especially true of things like weather, earthquakes, floods, and other natural phenomena. It is always a good idea to draw on this kind of data first if available.

RISK AND BASE VARIABILITY

Risk-based estimating uses two main groups of data: (1) base estimate and (2) risks. The previous chapters have discussed at large the base estimate (cost and schedule) and risks but with little emphasis placed on the magnitude of the risks that have to be included in risk analysis. Until now, the reader has learned that the number of variables that are included in analyses should be limited. Some authors recommend no more than 20 variables.[8] Other practitioners do not bother to limit them at all (we have seen 200 to 300 risks identified, quantified, and analyzed). We would like to have no more than 15 to 20 risks, realizing that every project and its risks are different.

The number of risks is directly connected to the magnitude of risks and their impacts for each project. Experienced cost risk analysts have clearly presented their position of defining the threshold of risks' impact value that qualifies them to be considered for risk analyses. The next paragraph cites a quote from Michael W. Curran (an AACE International Fellow) on his perspective on items that have to be included in a successful cost risk analysis ("RP" refers to AACEI Recommended Practice 41R-08):

> . . . the method for identifying critical items is strictly a top down drilling process guided by application of the criterion found in the critical variance matrix on Page 2 of the RP. This results in relatively few critical items—by design! Identifying critical items—by strict adherence to the critical variance/drill down procedure—yields relatively few such items (not to save time but to avoid error!). If the drill down goes too far south (thus violating the critical variance criterion) **many more items will be identified among which correlations likely exist, correlations either unknown or improperly understood.** One of the major purposes of R.E. is to "stay above the battle" so to speak. Don't get mired in detail. (I have seen estimators cringe when I told them this.) I am fully empathetic. I know what it's like to suppress primal urges. The world of physics (my training) is not that dissimilar to the world of estimating,

at least in one major respect—both feel compelled to omit nothing and measure everything! Any aspiring risk analyst doing that will no doubt be put on a different career path forthwith. (Cost-risk analysis, the form we see within AACE, involves a healthy dose of art as well as science. If the art takes on some semblance of logic and proves worthwhile over a sufficient number of applications, we can call it a ''heuristic.'' Range estimating, as described in the RP, is a heuristic.)[9]

The message is quite strong and clear—risk analysis is about analyzing events that significantly change the project cost or/and schedule when they occur. In our view the number of risks depends on project specifics. We consider that one of the first questions that a risk elicitor should ask about a project's risks is: What is keeping you awake at night? The answer to this question identifies a few risks and these risks are the most significant ones. Then the risks' elicitation goes down to lesser impact values and it must stop when the most likely impact value is about 75 percent of base variability. The factor of 0.75 has no scientific explanation but we considered that base variability includes all risks with impact value less than this factor. If we are continuing ''drilling down,'' as Michael W. Curran wrote, we may not only dilute the analysis resolution and complicate the model (sophistication) but we introduce errors in analysis by double counting the effect of lesser risks. The effect of double counting of lesser risks is like adding a moving contingency to the project that increases when the budget uses higher confidence levels of the cost range without having an explicit understanding of this contingency.

Large numbers of risks quantified and analyzed may complicate the project risk mesh. *Risk mesh* represents the project risk fabric and a sample of risk mesh is presented on Figure 4.1. Having many risks increases the manifold mesh and makes it difficult and unproductive for the

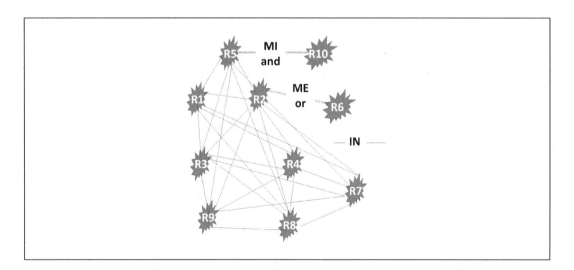

Figure 4.1   Risk Mesh—MI implies inclusion (and); ME implies exclusion (or); IN implies independent of each other

identification of the risks' conditionality. Risks' conditionality is the essential ingredient of RBE because together with model algorithm, they orchestrate the simulation.

Figure 4.1 shows a glimpse of the complication in establishing the dependency among risks that may be produced when the nonsignificant risks are quantified and analyzed. Each pair of risks must be tested for dependency and then for correlation. Testing risks for conditionality is not an option of risk analysis; it is a "must do it," since otherwise the results will not reflect what may happen. There are situations when three or more risks are in conditional relationship, and this may add more complexity to the risk mesh.

Figure 4.2 shows how risks contribute to the end result. Each independent event (IN) (R1, R2, R3, R4, and R7) is launched by YES/NO toggle using random number or other function such as: uniform or Bernoulli distribution. R10 may occur only when R5 occurs so the dependency may capture the scenarios of: (1) R5 occurs alone, (2) R5 and R10 occur simultaneously, and (3) none of them occurs. This dependency is called mutually inclusive (MI) and is realized by placing YES/NO toggle between R5 and R10.

Moreover, R6 may occur only when R2 does not occur. This represents an example of mutually exclusive (ME) risks. R6 may occur only when R2 does not so the dependency may capture the scenarios if: (1) R2 occurs alone, (2) R6 occurs only when R2 does not, and (3) none of them occurs.

Figure 4.3 describes how conditionality orchestrates the risks' dependency and correlation. The triangular shape symbolizes cohesion between: (1) risks' dependency, (2) risks' correlation,

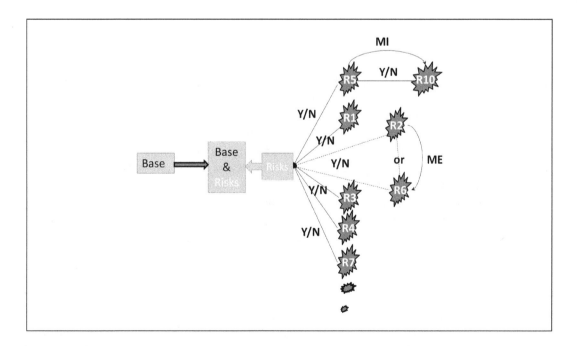

Figure 4.2   Events Dependency

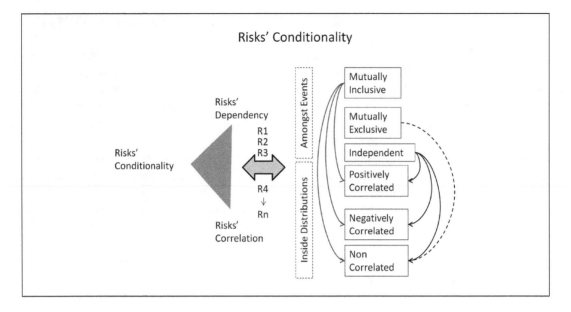

Figure 4.3   Risks' Conditionality

and (3) risks' conditionality. The risks' conditionality is tested on all risks quantified and the number of tests for *n* risks is:

$$C(n,r) = \frac{n!}{r!(n-r)!}$$

where: r = 2.

So if 15 risks are elicited the number of tests is 1,365. It is a large number, but don't worry, most of the time the test is done in fractions of seconds and our brain picks up quite rapidly whether or not a dependency exists.

Risks' dependency is checked first in a sense of how an event may affect or may be affected by other events. At the time a relationship is spotted the risk elicitor must focus on defining and documenting the type of dependency and then checking into correlation between risks' distribution. Figure 4.3 indicates that mutually inclusive and independent risks allow all types of correlations while mutually exclusive risks do not permit correlation.

Figure 4.3 shows that the dependency is related to the events—interrelationship, and correlation is related to how values are sampled from the inside distribution's impact.

RISK CONDITIONALITY

Figure 4.3 may be viewed as a checklist that the risk lead must go through to make sure all risks' characteristics are properly captured. Many times conditionality is evident and the process may take a few seconds but there are situations when establishing the right conditionality among risks takes careful consideration.

A general desire of any workshop participant is, "Make it short." The risks test of conditionality goes rapidly and smoothly when workshop participants understand the meaning of conditionality among risks and why it is important to capture it. The next several paragraphs will define and clarify the conditionality myth. The reader should be aware that the risk community is still using some of these terms interchangeably.

Our definition of risk conditionality relates to two main risk characteristics: (1) probability of occurrence and (2) impact (effect). Two categories of relationship between risks are possible: (1) dependency that is referring to probability of occurrence, and (2) correlation that is referring to how values are sampled from the distribution's impact.

Risks' Dependency

Risks' dependency involves the type of risks' conditionality that relates to the event causality. For example, risk R1 may occur only if risk R2 occurs. The dependency has three distinct scenarios: (1) mutually inclusive (MI), (2) mutually exclusive (ME), and (3) independent (IN). While independent risk is an easy concept (each risk has its own destiny), the other two require further discussion.

Mutually Inclusive

Mutually inclusive risks have two options: (1) total inclusiveness, which may be called "mutually inclusive 100 percent" (MI-100%), and (2) partial inclusiveness, which may be called "mutually inclusive P percent" (MI-P%.) Mutually inclusive 100 percent is presented in Figure 4.4, which depicts the situation when Event X has 50 percent probability of occurring, and Event Y *always and only* occurs when Event X occurs.

For risk management purposes, a legitimate question may be raised: Why not pack these two events together under one single risk? Condensing risks under a single one is recommended whenever there is reason to do so. Combining risks under a single risk is advisable, but the risk mesh needs to be well understood. There are situations when combining risks is not possible because they may affect different activities or, to assist with risk management, it is best to keep them separate. For example, a change in design may require additional construction cost (Event X) and additional land acquisition cost (Event Y). Each event is applied to distinct activities, which represent different project phases—schedule different, inflation different, different algorithm.

Figure 4.5 shows the situation when Event Y has 50 percent probability of occurrence that may occur only when Event X happens. This is an example of partial mutually inclusive (MI-50%.) Event X has also 50 percent probability of occurrence. The overall risk mesh is described by: 50 percent none of the events occur, 25 percent only base and Event X occur, and 25 percent base and both events occur.

Examples of mutually inclusive events:
1. Environmental regulatory agencies may require construction of additional mitigation ponds (likelihood = 30 percent). Additional mitigation ponds increase the cost of construction by: low = $2.5 M, most likely = $4 M, and high = $12 M: and because no land is available

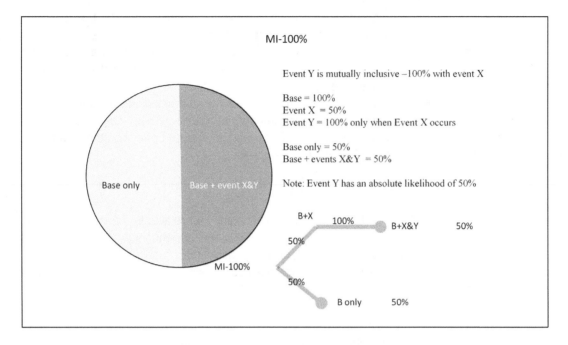

Figure 4.4   Mutually Inclusive—100 Percent Events

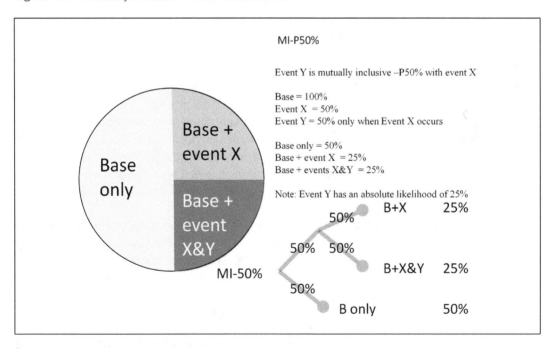

Figure 4.5   Mutually Inclusive—P50 Percent Events

for ponds construction, additional land acquisition (low = $0.5 M, most likely = $1 M, and high = $3 M) is needed. The example represents total mutually inclusive (MI-100%).

2. Environmental regulatory agencies may require construction of additional mitigation ponds (likelihood = 30 percent). Additional mitigation ponds increases the cost of construction by: low = $2.5 M, most likely = $4 M, and high = $12 M; and because the land available for ponds construction may not be enough, there is a 50 percent chance that additional land acquisition (low = $0.2 M, most likely = $.8 M, and high = $2 M) may be needed. This example represents partial mutually inclusive (MI-50%).

3. The building may need a deeper foundation (likelihood = 50 percent) and additional cost (low = $10 M, most likely = $20 M, and high = $50 M) may be needed. When a deeper foundation is constructed, the peripheral ground settlement may require additional ground support (likelihood = 30 percent) and have an impact of: low = $5 M, most likely = $8 M, and high = $15 M. This example represents partial mutually inclusive (MI-30%).

Another example of mutually inclusive risks is presented in Cartoon 4.1. The project is "indulging treats" and it requires the simple task of crossing the river. Risky (the dog's name) may think that there is 95 percent chance of having the delicious treats and just 5 percent that he may need to swim a little bit if he drops off the trunk. It looks tempting and Risky decides to go for it. Once Risky is on the trunk he discovers that if he falls off the tree trunk there is a 99 percent chance that he may be eaten. The project "indulging treats" may be affected by two events that lead to three outcomes: (1) getting the treats without problems [95 percent], (2) falling off the tree trunk and missing the crocodile teeth and getting the treats after a furious swim [.05 percent], and (3) getting caught by the crocodile [4.95 percent] and letting the treats spoil.

Cartoon 4.1 Mutually Inclusive Risks

Risky can be caught by the crocodile only if he drops off the tree trunk.

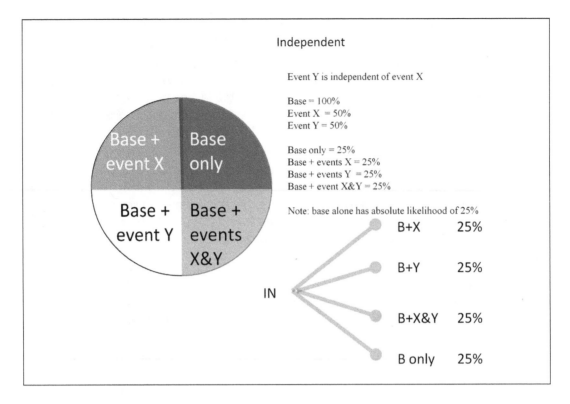

Figure 4.6   Independent Risks

Independent Events

Independent events—Let's have the same events as previously presented but now the events are independent (IN) of each other. Event X has a 50 percent probability of occurrence and Event Y has the 50 percent probability of occurrence and they are independent of each other. Figure 4.6 describes the configuration of risk mesh that is quite different than the previous one (MI-50%). By changing the risk dependency from partial mutually inclusive to independent risks, it changes the risk mesh to: 25 percent none of the events occur, 25 percent Event X occurs, 25 percent Event Y occurs, and 25 percent both events occur in addition to the base. It is quite different, isn't it? More about significance of these changes will be presented in the next chapters. The independent risks (IN) are so frequent that we will not provide any example of them.

Mutually Exclusive

A mutually exclusive situation has two options: (1) total exclusiveness, which may be called "mutually exclusive 100 percent" (ME-100%), and (2) partial mutually exclusive P percent (ME-P%). A mutually exclusive 100 percent relationship is presented in Figure 4.7, which depicts the situation when Event X has 50 percent probability of occurring, and Event Y *always and only* occurs when Event X does not occur. In this case, the base alone never occurs.

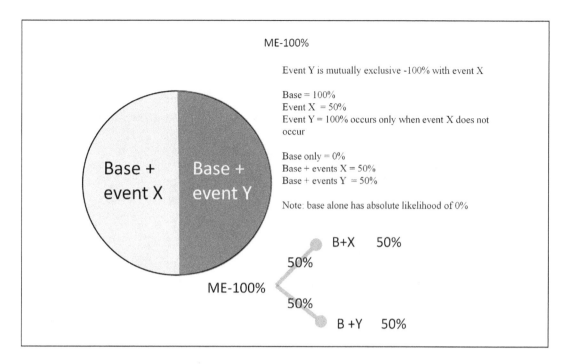

Figure 4.7   Total Mutually Exclusive Risks

More flexible mutually exclusive dependency is represented by partial mutually exclusive risks. Figure 4.8 presents the situation when Event X has 50 percent probability of occurrence and Event Y has 50 percent probability of occurrence only when Event X does not occur. It is the scenario of ME-50 percent.

Figure 4.8 presents a different risk mesh than what Figure 4.7 shows. The base alone occurs 25 percent of the times.

It is interesting to examine the particular case of "No Clue Events," when the risks and base alone have equal chances of occurrence. For this case, the risk mesh is defined by Event X with approximately 33 percent probability of occurrence (one third), and Event Y, which may occur only when Event X does not occur and has 50 percent probability of occurrence in a constrained condition. Figure 4.9 presents No Clue Events in graphic form.

Examples of mutually exclusive events:

1. A project stormwater drainage system *requires* additional stormwater storage capacity. The project team contemplates two options of meeting the project's requirements: (1) building stormwater vaults (likelihood = 50 percent) and construction cost (low = $2 M, most likely = $2.5 M, and high = $4 M) or acquiring additional land for diverting the stormwater (low = $1.5 M, most likely = $2.5 M, and high = $5 M). The example represents ME-100%, since the project must choose one or the other event.

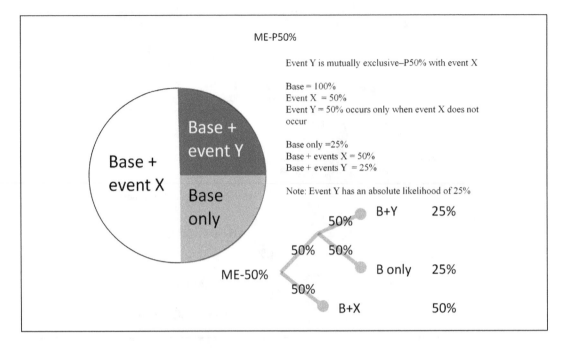

Figure 4.8   Partial Mutually Exclusive Risks

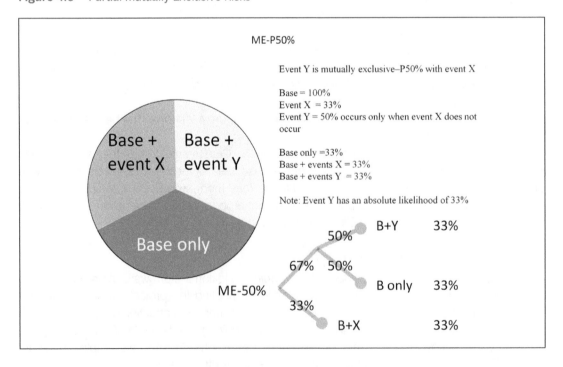

Figure 4.9   Mutually Exclusive—No Clue Events

2. A project stormwater drainage system *may require*: (1) building stormwater vaults (likelihood = 50 percent) with a construction cost (low = \$2 M, most likely = \$2.5 M, and high = \$4 M) or acquiring additional land (likelihood = 60 percent of remaining) for diverting the water surplus (low = \$1.5 M, most likely = \$2.5 M, and high = \$5 M). The example represents ME-60%, because the land acquisition may happen (likelihood = 60 percent of remaining) only when the additional stormwater vaults are not considered. The risk mesh is characterized by: 50 percent chance of building additional vaults, 30 percent chance of acquiring new land for diverting the stormwater, and 20 percent chance base only.

3. The classic example of mutually exclusive events is given by the definition of base uncertainty: (1) market worse than expected (likelihood = 10 percent) with increased cost of 20 percent, (2) market better than expected (likelihood = 30 percent) with decreased cost of 10 percent. These two events cannot occur simultaneously; they are in a mutually exclusive relationship. The risk mesh is characterized by: 10 percent chance of worse than expected market conditions, 30 percent chance of better than expected market conditions, and 60 percent chance of expected market conditions.

4. Table 3.5 shows one event that shall be in a mutually exclusive relationship with the other three events. The events "Earth Pressure Balance Tunneling Boring Machine," "Risk Sharing Procurement," and "Design-Builder Innovation" are substantial cost reductions that will avoid a "default on the contract." So all three events (opportunities) must be in mutually exclusive relationships with the threat event "contractor underbids." One way of creating the risk mesh of the four events presented here is as follows: contractor underbids event takes 10 percent probability of occurrence and the next three events are independent of each other but each of them is partial mutually exclusive event related to "contractor underbids." Figure 4.10 shows the risk mesh proposed configuration.

   The percentage for each plausible cause could be calculated using a rigorous mathematical formula, but why? We do not want to dig too deeply in statistics as long as we do not need to do it, and in this case, the risk mesh algorithm will do for us.

5. A business located in the project's proximity may require significant but not major accommodation and there is 40 percent likelihood that it will cost an additional: low = \$2 M, most likely = \$3 M, and high = \$4 M. In addition, there is 20 percent probability that the negotiation goes badly and the impact may be: low = \$8 M, most likely = \$10 M, and high = \$12 M. These two events may be captured in mutually exclusive dependency: the second event may occur only when the first event does not occur. In this case, the base alone has a likelihood of 40 percent occurrence. The partial mutually exclusive risk is defined as ME-33 percent. That means one-third of the times, when the first risk doesn't occur, the new risk occurs.

   There is a second way to capture this dependency: we may consider that the second event may occur only when the first event occurs by incorporating in its impact the consequences of the first event. Partial mutually inclusive dependency satisfies this new approach of risks' dependency. In this case, the dependency is defined by MI-50 percent

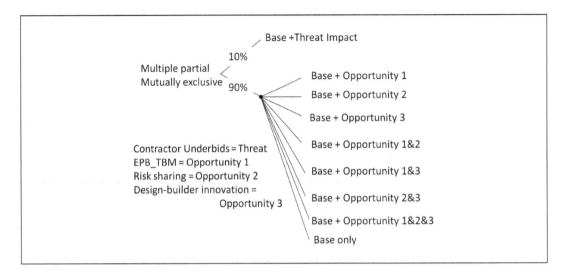

Figure 4.10 Risk Mesh of Three Independent Events and the Fourth Event Mutually Exclusive with All of Them

TABLE 4.2 Comparison between Two Methods of Capturing Risks' Dependency

Name	ME Mesh	MI Mesh	Delta
Minimum	2.04	2.04	0.00%
Maximum	15.38	15.18	1.28%
Mean	4.97	5.03	-1.04%
Std Deviation	4.02	4.05	-0.79%
Variance	16.13	16.39	-1.58%
Skewness	1.53	1.48	2.84%
Kurtosis	3.43	3.30	3.95%
Errors	0.00	0.00	
Mode	3.10	2.98	3.85%

Percentile	ME Mesh	MI Mesh	Delta
5%	2.41	2.42	-0.28%
15%	2.63	2.64	-0.12%
25%	2.79	2.80	-0.14%
35%	2.93	2.93	-0.15%
45%	3.07	3.07	-0.10%
55%	3.20	3.21	-0.15%
65%	3.36	3.37	-0.34%
75%	3.58	3.60	-0.53%
85%	12.37	12.40	-0.25%
95%	13.61	13.60	0.03%

and the impact is: low = ($8–$2) M, most likely = ($10–$3) M, and high = ($12–$4) M. The results of these two approaches should be close enough and acceptable. We would like to remind readers that it is essential that risk mesh is well understood by workshop participants. Remember:"I would rather be approximately right than precisely wrong."

Table 4.2 shows the percentile values of these two approaches of risk elicitations. It is noticeable that the results are quite close and the relative variance is insignificant for our level of analysis.

The previous example shows that dependency between risks may be captured using different methods—MI or ME. This practice is acceptable as long as it represents the project's risks' dependencies.

Another way to illustrate mutually exclusive risks is presented in Cartoon 4.2. Risky (our dog) wants to go home and nothing will stop him. At this moment there are three trails that lead to his home: trail 1 is straight through the tunnel, trail 2 is a long and tiring walk, and trail 3 involves crossing the river.

Cartoon 4.2   Mutually Exclusive Risks

Choosing one trail eliminates the other two trails. Risky cannot select two or all three of them. So trail options are in mutually exclusive relationship.

Schedule Risks

Schedule risks behave differently than cost risks. Cost risks, when they occur simultaneously, are simple to cumulate. Schedule risks require additional clarifications regarding their impact. The general assumption made about schedule risks is that they affect only critical activities (activities that belong to the critical path). In other words, a schedule risk indicates that the critical path schedule is going to be affected. The project's flowchart diagram facilitates the schedule risk analysis because it creates the algorithm that defines critical path.

When two or more schedule risks occur on the same activities, the flowchart diagram cannot assess how to handle them. In this case there are two major situations: (1) schedule risks are in parallel, and (2) schedule risks are in series. Figure 4.11 depicts the situations when the same schedule risks applied on the same activities but they are in series or in parallel relationship. It is clear that the results are significantly different since, when risks are in parallel, the model selects the higher risk value and when risks are in series the model calculates the sum of them. Further details will be presented in Chapter 6.

When schedule risks affect parallel tasks of the same activity it means that one task delay does not affect the duration of the other task and, as a result, their cumulative impact on the activity's duration is given by the longest delay risks. In other words, if an activity is affected by many parallel duration risks, the dominant schedule risk decides about the activity's delay.

The series schedule risks are applied on tasks in series of the same activity so if one risk delays the first task, that delay is transferred to the next task, which is affected by its own schedule risk. In this case, the cumulative effect of these two schedule risks is calculated by adding the delay produced by both risks.

Risks' Correlation

Risks' correlation is the type of risks' conditionality that goes inside of risks' distribution impact. Correlation indicates how the values from inside risk impact distributions are sampled. Theory about correlation between two or more variables may be complex. Just Google the term and

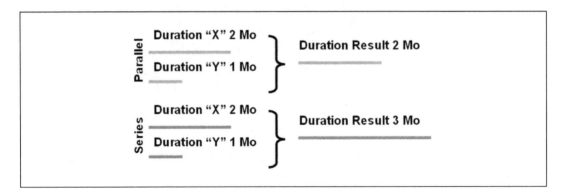

Figure 4.11   Schedule Risks—Series versus Parallel

you will be overwhelmed with information. The authors do not intend to elaborate on correlation definition or meaning. We are presenting the pragmatic consequences of correlation and its effect on risk-based estimating.

Figure 4.3 indicates the assumptions we made to reflect the pragmatic side of correlation. It shows that correlation may be applied only on risks that occur on the same iteration (realization) during simulation. For example, when risks are in mutually exclusive relationship the correlation can't exist. Correlation has weak meaning when risks are independent, and in that case correlation has effect only when both risks occur on the same iteration. In the case of mutually inclusive risks the correlation effect is substantial. Chapter 6 will illustrate in detail the correlation effect on risk analysis results.

Now let's assume that we have two risks that occur at the same time. Risk X, whose distribution impact ranges from 1 to 10, and risk Y, whose distribution impact ranges from 4 to 22. The distributions may be symmetrical or nonsymmetrical. These two risks may have three forms of correlation: (1) positive correlation, (2) negative correlation, and (3) noncorrelation.

Positive Correlated Risks

Positive correlation between these two risks indicates that the affected risks move in the same direction. If one risk takes a value from the high end area of its range the second risk takes a value from its own high end area of its range, too. In other words, if one risk goes high, the other risk goes high too for the same iteration. Positive correlation is presented in Figure 4.12.

Figure 4.12 indicates that if risk X goes low, risk Y goes low, too, and if risk X takes a value around mean, risk Y takes a value around mean as well. Examples of positive correlated risks are presented next.

A classic commonly used example is the correlation between increase of cost and increase of duration delay when additional construction effort is needed. This case is a little different because it represents the same risk with two impacts: (1) cost impact and (2) schedule impact. It may be considered as two total mutually inclusive events.

The examples offered when we have illustrated mutually inclusive risks may very well be complemented with association of positive correlation. It may be the case when the construction cost is proportional with the area of land acquired. Assuming that the land cost depends on the magnitude of the acreage acquired, then a positive correlation between land acquisition cost and construction cost is warranted.

If positive correlation is assigned to those two risk events, the second event has no choice of sampling its own value. Let's consider the first example: Environmental regulatory agencies may require construction of additional mitigation ponds (likelihood = 30 percent). Additional mitigation ponds increase the cost of construction by (low = $2.5 M, most likely = $4 M, and high = $12 M) and because no land is available for ponds' construction, additional land acquisition (low = $0.5 M, most likely = $1 M, and high = $3 M) is needed. If the first risk event takes a value of 11.5, the second risk has no other choice than to have the value of 2.89; if the first risk takes a value of 2.6, the second risk will take .54; if the first risk takes 4.83, the second risk takes 1.19, and so on.

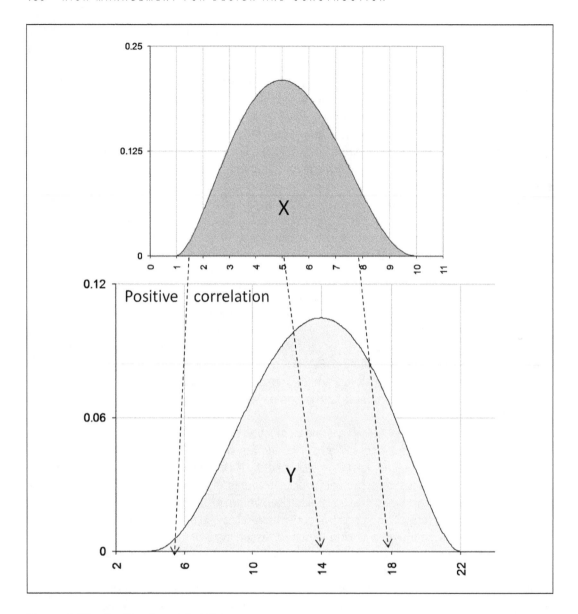

Figure 4.12 Positive Correlated Risks

It is important that each risk distribution be well understood (What is included in low value? What is included in high value? What is included in most likely value?) in order to decide about correlation that may exist between distributions.

Positive correlation may be applied to more than a pair of two risks. Sometimes it is used to represent the conditionality of uncertainty of multiple base values. The authors do not recommend this approach because it encourages fragmentation of cost and/or schedule, excessive noise in analysis, and so on, which leads to *professional sophistication.*

Positive Correlated Base Uncertainties

Some risk professionals use positive correlation among many base cost uncertainties to overcome the narrowing effect of having many variables in analysis. Having many variables creates the illusion of detail analysis (professional sophistication) but if the positive correlation is not employed, the results are in an unacceptably narrow range. Positive correlation among multiple base uncertainties eliminates the narrowing effect on the total base uncertainty but introduces two major fallacies: (1) the correlation may not be justified by project conditions (correlations should not be applied as a blanket and positive correlation should not be used as a tool to increase the analysis range—no matter how tempting it is), (2) the noise introduced in analysis results raises the question: "Why are we doing RBE?"

The Problem

The second fallacy is the most disturbing one because its detrimental effect is not so obvious. Chapter 3 has presented the noise induced in RBE results by base variability. The Chapter 3 showed that large values for base variability are detrimental to the risk-based estimating process. The positive correlation between base variability distributions amplifies the noise detrimental effect by making the noise so loud that the risks become mute. When risks cannot speak for themselves, then the question "Why are we doing RBE?" is legitimate and we will ask it no matter what.

"Why are we doing RBE?" is a crucial question that leads to a credibility issue when someone is asking it after receiving risk analysis results. When positive correlation is indiscriminately used and analysis results look like perfect S curves, you and I will ask that question. Let's see your reaction, as the owner of a large project, after learning about a recent RBE results. The project has a base cost of $1.12 B and the baseline schedule ends four years from now. The project's flowchart is presented in Figure 4.13, and it looks quite complex at best.

The overall project cost is represented in Figure 4.14. You and I may be troubled by what Figure 4.14 shows and we may ask: Why are we doing RBE? We are noticing that risks are quite underrepresented because: (1) risks add to the project $7 M (0.74 percent of nonrisks' value) at 10 percent probability of not exceeding, (2) risks add to the project $15 M (1.24 percent of nonrisks' value) at 50 percent probability of not exceeding, and (3) risks add to the project $27 M (1.83 percent of nonrisks' value) at 90 percent probability of not exceeding. These results make me, and perhaps you, think: Next time I will apply a blank range over the base value and not spend money on risk analysis. We may remind ourselves that RBE is beneficial because it provides data for risk management.

The risk team has used the multiple base uncertainties approach that introduces 21 wide-range distributions. If these distributions are independent, the overall base uncertainty will have an unacceptable narrow range (see Chapter 3). So positive correlation has been applied on all of the 21 base uncertainties and the total base uncertainty has become quite large. The

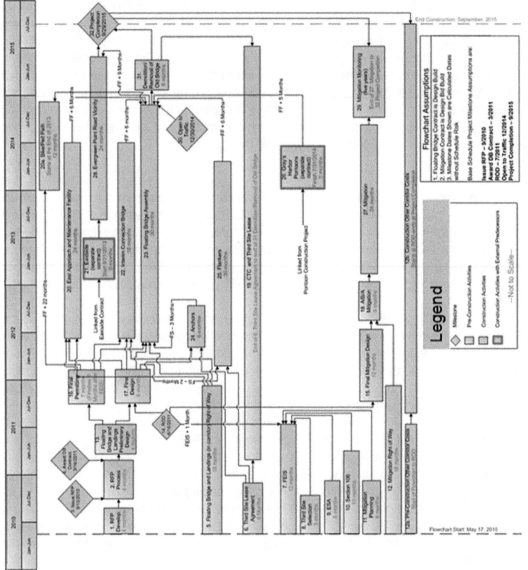

Figure 4.13 Project Flowchart Diagram

138

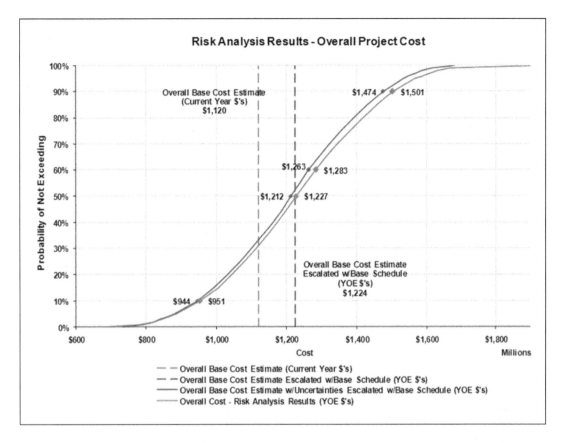

Risk Analysis Results - Overall Project Cost

Figure 4.14   Total Project Cost

process introduces two major fallacies: (1) indiscriminate use of correlation and (2) very large base uncertainties.

Conditionality (dependency and correlation) represents relationships that exist among variables. The risk analyst must apply correlation between two variables only if variables behave in that way. But the model applies positive correlation to all variables regardless of what they represent. Only this procedure alone reduces or terminates the effectiveness of the RBE process.

Table 4.3 shows the values of each base uncertainty and the total base uncertainty obtained by using positive correlation among all base uncertainties. A robust cost risk analysis requires that each of the 21 base uncertainties be tested against correlation and findings be documented and applied to a correlation (positive or negative) only when it is warranted. A simple examination of Table 4.2 leads to questions such as: How is activity Mitigation Right of Way positively correlated with Final Design or Demolition/Removal of Old Bridge? The questions may continue on and on.

TABLE 4.3   Base Cost Uncertainties

Activity	Project Base Cost Uncertainty Ranges by Activity		
	10th Percentile	Base Estimate	90th Percentile
RFP Development	$2,129,994	$2,662,493	$3,194,991
RFP Process	$12,539,976	$16,649,970	$18,779,964
Mitigation Right of Way	$13,786,833	$18,193,333	$20,638,167
Pre-Construction Other Corridor Costs	$11,035,421	$17,731,776	$24,428,131
Construction Other Corridor Costs	$49,288,257	$70,797,821	$92,307,386
~~ing Bridge and Land~~ ~~sign~~		$35,355,855	
A/B/A ~~n.~~	$283,999		$425,999
CTC and Third Site Lease	$23,454,138	$32,363,054	$36,363,054
East Approach and Maintenance Facility	$56,193,012	$72,937,228	$88,109,307
Bike/Ped Path	$16,084,968	$22,241,307	$29,340,570
Interim Connection Bridge	$22,139,457	$27,065,814	$34,595,600
Floating Bridge Assembly	$367,692,544	$458,117,004	$540,229,533
Anchors	$12,595,398	$15,961,631	$18,135,194
Flankers	$168,895,506	$220,515,254	$253,474,205
Demolition/Removal of Old Bridge	$20,641,804	$25,853,939	$31,437,690
Total	$860,517,658	$1,120,067,808	$1,349,202,255

The fact that the low and high values of each distribution represent the 10 percent and 90 percent of distribution range values contributes to the final large range because 20 percent of the times the model picks up values from either outer low or high limit zone.

Having a large range for base uncertainty (reality is that what the project calls base uncertainty is in fact base variability, as we presented in Chapter 3) creates intense and widespread noise in analysis and it hides the risk's effect, whatever it may be. Once again, we are advising our readers about the importance of balancing the base uncertainty with risks' magnitude. When the base uncertainty is too large, the meaning of RBE is lost, and with it, the usefulness and credibility of cost risk analysis is at risk.

The Solution

The authors have communicated with the cost lead of the project base cost review team and expressed concerns about the cost risk analysis results and the effect of having large and positive correlated base cost uncertainties. The base cost lead sent us his understanding and interpretation of the base cost review process. The following paragraphs present the base cost lead email with our comments interpolated in the original text using italics font.

Regarding uncertainty and market conditions:
I really like how you handle the market conditions issue with your table. This is how the RBE team and SMEs rate the market conditions *in the future as changed from today*. I think it captures the potential change to the bidding climate for change

in competition, general economic cycle, and commodity price changes between now and the time of performance. This is a partially manageable issue. *This is okay.*

The uncertainty range around the base cost is different. The 10 percent or 90 percent range just indicates that we are uncertain of the exact price *if bid today.* Our base estimate for time and cost is what we think will happen if we let today. One out of ten times, the price could be very low or very high, today. *Correction: Two out of ten times the cost could exceed the estimated low or high values. Only 80 percent of times the model will select values from inside the range presented.*

We do not expect the limit of the range to happen; we expect the base cost to happen (the mode). *The values around most likely have higher chance to occur than values from tails area.*

If the range is asymmetrical, it indicated that our RBE team and SMEs have noted that in today's market (not market condition change) there may be more uncertainty in one direction or another from the base estimate. We could change the base estimate to the mean; then all would be symmetrical.

Now we need to discuss in detail.

Let's take the Flankers' (cost item on Table 4.3) base cost and uncertainty as an example. The base uncertainty provided by the base cost team is defined by a Trigen distribution function with the bottom representing 10 percent and top representing 90 percent. The bottom is about $169 M, most likely (mode) is about $221 M, and the top is about $253 M. The distribution is shown in Figure 4.15. (By the way, we try to avoid this distribution because in many cases the SMEs do not understand its consequences and may create unwanted results if the distribution is not presented during elicitation, as we have presented in Chapter 3.)

The graph indicates that the Flankers' cost estimate ranges from a minimum value of approximately $132 M to a maximum value of about $285 M. We must agree that it is a large range that requires further analysis since we need to identify and quantify the drivers that push the cost to its ends. Figure 4.15 shows that there is 10 percent probability that Flankers may cost between approximately $132 M and about $168 M and another 10 percent probability that the Flankers may cost between $253 M and $285 M. Significant changes affecting the Flankers' cost may happen and we need to identify and quantify them (in other words, risk assessment). A simple large blank range is not acceptable. We need to tell a story why the low cost can be so low and why the high cost can be so high.

To bring this issue in line with the spirit of cost risk analysis, the following solution may be used: Base cost of Flankers is about $221 M with its variability ±5 percent. The Trigen distribution representing base variability is defined as: low of approximately $210 M, a most likely of about $221 M, and a high of about $232 M. The low represents 10 percent and the high represents 90 percent. It is sophisticated, isn't it? I am trying to stay on project general algorithm assumptions. (Personally, I recommend the usage of minimum, most likely, and maximum when a distribution is defined.)

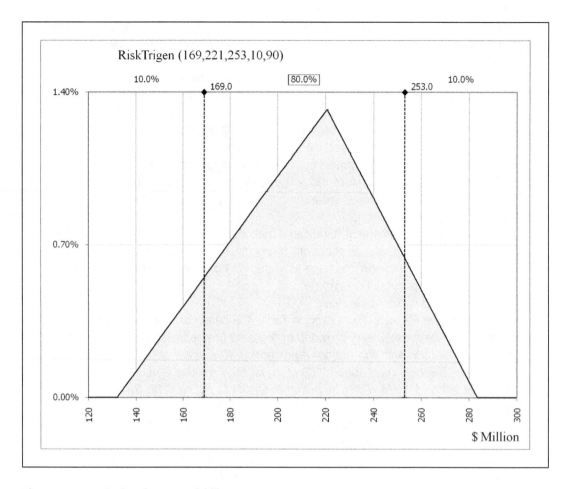

Figure 4.15 Flankers' Base Variability

Now we need to capture the events that bring the cost down or up. Whatever they are, we need to have a clear description of these events. The probability and the impact of these two events that are presented next (they are examples only) are designed to match in a reasonable fashion the initial distribution range of base uncertainty provided by the base cost team. We may consider an opportunity with 35 percent probability to occur and an impact (savings) of: low $20 M, most likely $45 M, and high $65 M. Then we may consider a threat (mutually exclusive with the opportunity) with 20 percent probability of occurrence and an impact of: low $20 M, most likely $30 M, and high $50 M. The cumulative effect of this combination overlaid with the original base uncertainty is presented in Figure 4.16.

Figure 4.16 indicates clearly why the range is so large. Readers can see that something is going on because of three humps on the histogram. The left hump is

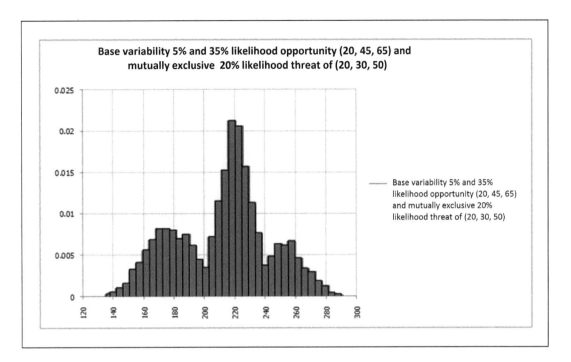

Figure 4.16   Combined Effect of Base Variability and Risks

created by the opportunity, the middle hump shows the as planned situation, and the hump on the right indicates the estimated cost when the threat occurs. The distributions presented in Figures 4.15 and 4.16 have approximately the same range but differ as to shape. At this time, nobody can say which of these shapes describes better the Flankers' uncertainty.

Furthermore, Figure 4.17 presents the cumulative distribution function for these two scenarios.

Figure 4.17 shows that the combined effect of base variability and risks is close enough to the simple base uncertainty distribution. In summary, the initial uncertainty must be replaced by base variability and two risk events because a large blank and asymmetrical base uncertainty defeats the purpose of RBE. The market condition will come on top of them.

When the SMEs say there is a 20 or 33 percent uncertainty in a particular element, they are not saying they expect the number to change that much, only that in rare occurrences it could be that much. The 10 to 90 percent range is a tool to get to a median (which is different than our base cost estimate—the mode) and to bracket the likely result today.

The 10 to 90 percent range hides the real minimum and maximum of the distribution range. In many cases, the SMEs do not realize how low or how high the distribution

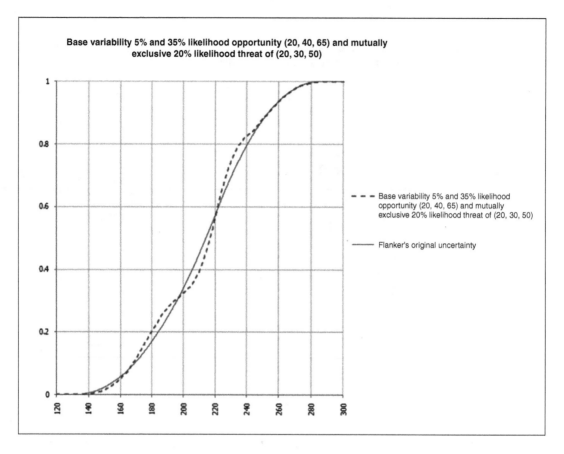

Figure 4.17   Comparision between Simple Asymmetrical Base Variability and the Combined Effect of Base Variability and Risks

> *goes. This is the reason I do not recommend usage of percentiles for low and high as distribution limits. The median and any other statistics measurements are defined as well by distribution minimum, most likely and maximum values.*

The solution presented in the previous paragraphs is just one among many others that the cost lead, risk lead, and SMEs may provide. The workshop leads (risk elicitor and cost reviewer) must be aware of unintended consequences of any change or simplifications that they might be tempted to make. The authors would like to remind readers that RBE focuses on estimates and risk assessment and both of them create data of equal values.

Negative Correlated Risks

Negative correlation between two risks indicates that the affected risks move in opposite directions (Figure 4.18). If one risk takes a value from the high end area of its own range, the

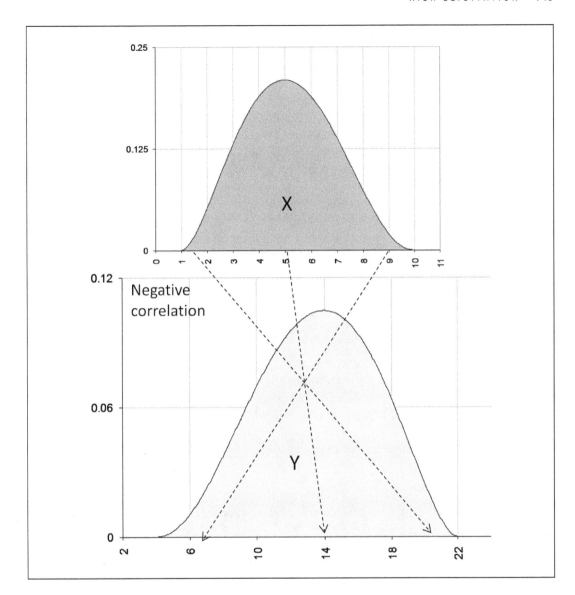

Figure 4.18   Negative Correlated Risks

second risk takes a value from its low end area. In other words, if one risk goes high, the second risk goes low. Negative correlation must be applied to the pair of risks. Applying negative correlation to multiple risks is difficult, if not impossible, and the risk elicitor has to analyze each particular situation and observe the implication created by correlation overall.

A typical example of negative correlation is represented by the so-called *crashing the schedule* procedure, which is used when meeting the project deadline is most important. Let's assume that there is an event that occurs during construction that has two impacts: (1) increases the cost (low = $2 M, most likely = $4 M, and high = $10 M) and (2) produces a delay (low = 4 Mo, most likely = 6 Mo, and high = 12 Mo). The project manager may assign more resources to deal with the event in order to reduce the delay so when the cost is high, the delay is low.

Let's consider again the first example of mutually inclusive risks: environmental regulatory agencies may require construction of additional mitigation ponds (likelihood = 30 percent). Additional mitigation ponds increase the cost of construction by (low = $2.5 M, most likely = $4 M, and high = $12 M) and because no land is available for ponds' construction, additional land acquisition (low = $0.5 M, most likely = $1 M, and high = $3 M) is needed. In this case, the SMEs consider that if more money is spent on construction, less money should be spent on land acquisition.

That means that a negative correlation between these two distributions is necessary. The negative correlation will dictate the way values are sampled during simulation. For the same iteration, if the first risk event takes a value of 11.5, the second risk has no other choice than to have the value of .54; if the first risk takes a value of 2.6, the second risk will take 2.89; if the first risk takes 4.83, the second risk takes 1.19, and so on.

Noncorrelated Risks

The most frequent relationship between risk distributions is the noncorrelation. In this case, the values sampled from each distribution depend only on the distribution type and random number. Each distribution has its own destiny. Figure 4.19 represents graphically the noncorrelated distributions. The second distribution may select any value from its range regardless of the value selected by the first distribution.

The noncorrelated distributions are default distributions of any software programs that assist in Monte Carlo simulation.

SUMMARY

Risks are a critical part of risk-based estimating, and their impact is more noticeable at the edges of distributions that represent a project's cost or schedule. Risk analysis is about what may happen at the end tails of distributions. So it is important to have an effective risk assessment in order to provide reliable information for the purposes of: (1) having a quality estimate for the cost and schedule, (2) managing risks, and (3) establishing the right budget and schedule for the project.

The quality of the risk assessment depends on the knowledge of the SMEs and the methods used during the risk elicitation process. The advance risks elicitation interviews of small groups of SMEs is the recommended method. It may take longer but it will provide better qualitative and quantitative data.

Any analysis or model is only as good as the data that inform it. Therefore, it is important to be aware of the effect of cognitive biases on the risk elicitation process. The risk elicitor must make an effort to become familiar with these and adopt strategies during the elicitation process to help minimize their effect on the quality of the data.

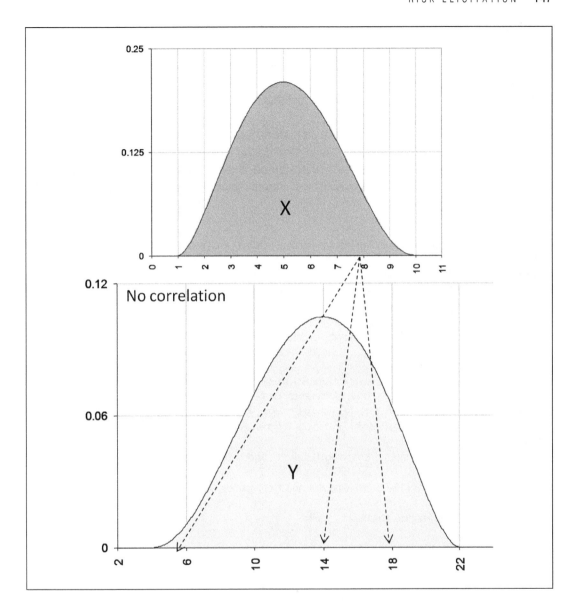

Figure 4.19 Noncorrelated Risks

"What is keeping you awake at night?" is the most important question that any risk elicitor must ask. Similarly, a second question should be asked: "What would you change about the project if you were the ruler for a day?" Asking these questions will help to expose critical threats and opportunities. Once these critical risks have been addressed, risk elicitation may proceed to the smaller risks and it should stop when the most likely impact value is about 75 percent of the base variability, which is a good general rule of thumb.

Establishing the right conditionality (dependency and correlation) among risks is essential for ensuring quality results from the model. If risks' conditionality is not properly captured, it is likely that the analysis will be flawed and produce misleading results. The conditionality has two major branches: (1) dependency that relates to the relationships among events, and (2) correlation that relates to how values are sampled from distributions when risks occur. Both branches have equal weight in a good risk assessment.

Replacing risks with large base variability is bad practice since it defeats the purpose of risk-based estimating. RBE assesses risks in order to manage them. In the rare situations when a large base variability appears correct, this approach is still detrimental to the entire effort because in essence it conceals risk events and hinders the ability to manage project uncertainty.

ENDNOTES

1. R.M. Hogart, "Cognitive Processes and the Assessment of Subjective Probability Distributions." *Journal of the American Statistical Association*, 70 (1975), pp. 271–294.

2. Amos Tversky and Daniel Kahneman, "Judgment under Uncertainty: Heuristics and Biases." *Science* vol. 185, no. 4157 (September 1974), pp. 1124–1131.

3. There are currently 45 African nations out of a total of 191 member states of the United Nations. This gives a probability of about 24 percent that any one UN member nation selected at random is located in Africa.

4. M. Alpert and H. Raiffa, "A Progress Report on the Training of Probability Assessors." In D. Kahneman, P. Slovic, and A. Tversky, eds. *Judgment Under Uncertainty: Heuristics and Biases* (New York: Cambridge University Press, 1982).

5. M.E. Hynes and E.K. Vanmarke, Reliability of Embankment Performance Predictions. Proceedings of the ASCE Engineering Mechanics Division Specialty Conference. (Waterloo, Ontario: Univ. of Waterloo Press 1976).

6. J.J.J. Christensen-Szalanski and J. B. Bushyhead, "Physicians' Use of Probabilistic Information in a Real Clinical Setting," *Journal of Experimental Psychology: Human Perception and Performance,* vol. 7 (1981), pp. 928–935.

7. T.S. Walsten and D.V. Budescu, "Encoding Subjective Probabilities: A Psychological and Psychometric Review," *Management Science*, vol. 29 (1983), pp. 151–173.

8. K.K. Humphreys et al. RP 41R-08 Risk Analysis and Contingency Determination Using Range Estimating, AACEI, 2008.

9. AACEI Recommended Practice 41R-08.

CHAPTER 5

RISK MANAGEMENT

> There are risks and costs to a program of action. But they are far less than the
> long-range risks and costs of comfortable inaction.
>
> —*John F. Kennedy*

INTRODUCTION

Every project faces its own unique challenges. Furthermore, every project is faced with limited resources, be they time, money, or workerpower. The project manager is faced with many choices concerning the allocation of these resources. As the management of risk is just one of the many activities that must be addressed, a decision will need to be made as to what level of risk management is needed.

In Chapter 1 of this book, the Project Management Institute's definition of *risk management* was introduced. The five steps identified in the PMI® definition are often presented in the form of a pyramid with the first step, risk management planning, at the bottom and the last step, risk monitoring and control, at the pinnacle. At first glance, the metaphor of a pyramid seems appropriate as the steps begin broadly and, as one climbs up, the focus narrows from the identification and analysis of risks, to the development of responses and actions that are then monitored through execution and completion of the project.

While the logic of this metaphor makes good sense from the perspective of process, it is useful to literally flip it upside down if we think about risk management from the perspective of what requires the greatest level of effort (see Figure 5.1). If the size of the pyramid's steps are assumed to represent the measurement of efforts needed to accomplish them, then each step increases as we move forward in the process. In other words, the best risk planning and analysis in the world is all for naught if the appropriate responses are not developed and implemented in a timely and effective manner. This point may seem trivial at first, but it is the experience of

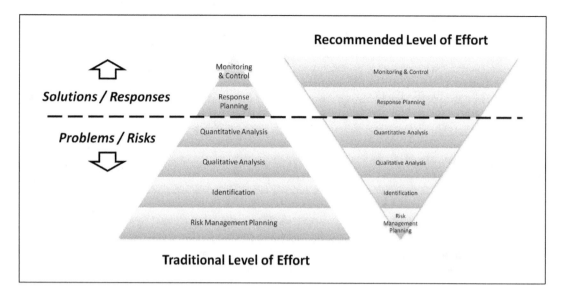

Figure 5.1   Traditional versus Recommended Emphasis of Risk Management

the authors that "traditional" risk management places more emphasis on problems and their analysis than on solutions and their execution.

Of course, effective risk management requires a balanced approach with respect to risks and their responses. What the authors wish to emphasize here is that effective risk management requires more than analysis—it demands decisive action. Risks require appropriate responses just as problems require viable solutions. Risk management is a dynamic process that requires the active participation of people to be successful. This chapter will discuss each step of the risk management process in greater depth. We will focus here on the "solution" side of the pyramid as the previous chapters have presented and discussed the "problem" side.

What Is My Project's Tolerance for Risk?

You may be asking yourself, "How much confidence am I willing to pay for?" In other words, what is the project's tolerance for risk? This is an excellent question that deserves some discussion.

From a mathematical perspective, it stands to reason that all projects should maintain a tolerance for risk at the 50 percent mark. For example, assume that a quantitative risk model prepared for a middle school project identifies a figure of $32.5 million as the median predicted project value. If there are no extenuating factors (as described in the following paragraphs), then this would be a reasonable level of risk to assume.

Sadly, we do not live in a perfect world. In light of this, a project's tolerance for risk will depend on a number of factors. These include:

- **Political sensitivity**—Projects that are highly visible in the public eye sometimes have a lower tolerance for risk. On such projects, there may be greater political will, which translates into a greater availability of project resources to mitigate risk. Also, the fear of failure may be more acute. Often, the sunk cost effect comes into play (which will be discussed in greater detail later in this chapter).

■ **Funding availability**—The availability of project funding can play a huge role in how risk is managed. There may not be funding available to mitigate risks as proactively as desired. In such cases, the tolerance for risk may be higher and not necessarily by choice. The performance of a quantitative risk analysis can provide the data necessary to paint a convincing argument to decision makers that additional funds should be sought now to avoid the expenditure of even more funds later down the road. To quote Albert Einstein: "There is no force more powerful in the universe than compound interest."

■ **Schedule criticality**— Often the project schedule will have a firm hand in determining a project's tolerance for risk. If there are key project milestones that must be met, this can increase costs dramatically, provided, of course, funds are available to be allocated. It has been the authors' experience that many of the so-called *critical milestones* identified in a project's schedule are completely artificial. Many are tied to erroneous statements made by policy makers early on in project planning. Again, a solid quantitative risk analysis can provide the data to show the cost associated with setting unreasonable milestones and can quickly change minds about what is truly critical and what is not.

Prospect Theory and Risk Tolerance

The framing effect is an essential component of *prospect theory*, which was first presented in 1979 by Kahneman and Tversky. Contrary to *expected utility theory*, which posits that people make decisions based on maximizing utility, prospect theory suggests that most people make decisions based on how prospects are framed. Consider the following situation.

Imagine that the United States is preparing for the outbreak of a deadly strain of influenza that is expected to kill 60,000 people. Two alternative programs to combat the disease have been proposed. Assume that the exact scientific estimates of the consequences of the programs are as follows:

■ If program A is adopted, 20,000 people will be saved.

■ If program B is adopted, there is a one-third probability that 60,000 people will be saved and a two-thirds probability that no people will be saved.

Which of the two programs would you choose?

Assume that two alternative programs to combat the disease have been proposed. Assume that the exact scientific estimates of the consequences of the programs are as follows:

■ If program C is adopted, 40,000 people will die.

■ If program D is adopted, there is a 33 percent probability that nobody will die and a 66 percent probability that all 60,000 people will die.

Which of these two programs would you choose?

A scientific study was conducted by Kahneman and Tversky posing similar questions to people. The results were quite remarkable:

- If program A is adopted, 20,000 people will be saved (72 percent).
- If program B is adopted, there is a one-third probability that 60,000 people will be saved and a two-thirds probability that no people will be saved (28 percent).
- If program C is adopted, 40,000 people will die (22 percent).
- If program D is adopted, there is a one-third probability that nobody will die and a two-thirds probability that 60,000 people will die (78 percent).

The results of Kahneman's and Tversky's study were fascinating.[1] Mathematically speaking, programs A and C are identical while programs B and D are identical. However, the majority of people selected programs A and D, which is completely illogical from a mathematical standpoint. What could have caused this breakdown in logic?

The explanation has to do with the manner in which the prospective outcomes are framed. The implicit reference point of the question is that if no program is adopted, then 60,000 people will die. The outcomes of programs A and B are stated in terms of gains, and people tend to be risk-averse when encountering opportunities for gain—respondents typically prefer to take the known outcome rather than the gamble.

The outcomes of programs C and D are stated in terms of losses, and people tend to become risk-seeking when confronted by the likelihood of loss—respondents will typically prefer the gamble over the certainty of losses.

Prospect theory postulates that people evaluate from the perspective of the *status quo* suggested by the way a prospect is stated, and think of each prospect as involving a gain, a neutral outcome, or a loss. The influence on decision making by the way in which the problem is asked or stated is called a *framing* effect, and can lead to irrational decision making and, consequently, poor value.

Figure 5.2 illustrates how the framing of a prospect impacts the value that is placed on the utility of expected outcomes.

The status quo serves as the reference point, which is indicated on the graph as Point A. Many choices involve decisions between retaining the status quo and accepting an alternative to it. Because prospects are evaluated in relation to the status quo, gains will be evaluated cautiously from a risk-averse point of view, and losses will be evaluated in a risk-seeking manner.

At Point B, further losses do not lead to a large decrease in value; however, comparable gains lead to a large increase in value. The person at Point B will risk small losses in order to obtain potentially large gains. This predisposition is referred to as the *sunk cost effect*. The sunk cost effect has two key aspects to it. First, people tend to have an overly optimistic probability bias, whereby after making an initial investment, the perception of that investment paying dividends is increased. Second, sunk costs appear to operate chiefly in those who feel personal responsibility for the investments that are to be viewed as sunk.

The sunk cost effect is often witnessed in the domain of public projects. Projects whose costs spiral wildly out of control are effective examples of this, where public officials refuse to

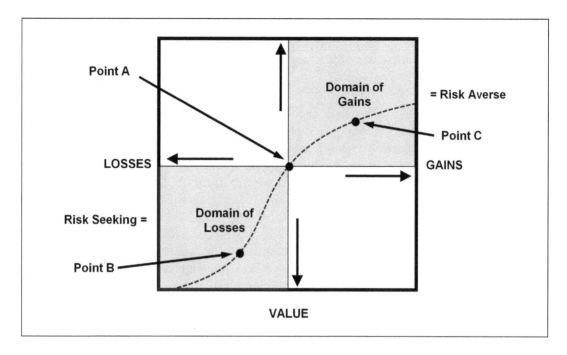

Figure 5.2   The Effect of Prospect Framing on Risk Perception

cancel a project due to the large financial and political investments made, even though doing so would offer much better value to the public welfare.

Conversely, a person at Point C will be reluctant to risk even small losses for large gains. This is because losses tend to loom larger than gains, and therefore a decision maker will be biased in favor of retaining the status quo. This is termed the *endowment effect*—it explains the reluctance of people to part with assets that belong to their "endowment."

The sunk cost and endowment effects can bias how risks are perceived based on the current status of a project at the time of risk elicitation and can therefore influence the responses of participants. A project that is significantly behind schedule or over budget may bias the information elicited from participants, just as can a project that is significantly ahead of schedule or under budget.

RISK RESPONSE PLANNING

If we are ignorant of a project's risks, then it is likely we will stumble upon them in the dark. Understanding the risks that have been identified allows the adoption of the proper framework for asking the right questions to address them. Throughout this book, the focus has been on enhancing our understanding of risk by identifying, assessing, and modeling risks through the process of developing a risk-based estimate. In this section, the focus will be on developing strategies to reduce the effect of these risks.

"If I had an hour to solve a problem and my life depended on the answer, I would spend the first 55 minutes figuring out the proper questions to ask. For if I knew the proper questions, I could solve the problem in less than 5 minutes."

—Albert Einstein

The quote above speaks to the importance of understanding the nature of a problem before solving it. Everyone is aware of the genius of Albert Einstein; however, much of his success can no doubt be attributed to his keen powers of perception and observation. If we express Einstein's breakdown of problem solving, he suggests spending about 90 percent of our time studying the problem and only 10 percent developing a solution.

With respect to risk workshops, the authors would tend to agree that this distribution of time is pretty accurate when risk response planning is included. The point must be made that many risk workshops stop once the analysis has been performed and leave the subsequent steps of risk management up to the project team to contend with. This is akin to a physician diagnosing a patient's ailments and then sending him out the door without a remedy. As the bewildered patient walks out the door, the doctor tells him, "Don't worry, I'm confident you'll find a cure!" If you were the patient under such a scenario, no doubt you would be both worried and upset.

The authors feel it is of vital importance that a risk workshop be comprehensive in nature and offer up potential solutions to the myriad problems they have identified and analyzed. This is the domain of risk response planning.

The process of risk response planning should address the identified and quantified risk events that could affect the project. The focus is on minimizing threats and maximizing opportunities. Risk response planning can occur as part of the primary risk workshop or scheduled as a follow-up workshop. Either way, it is essential that it be performed in a timely and planned manner.

The identification of risk responses requires an organized approach and methodology in order to provide the best results. The process can be broken down into the following steps:

- **Brainstorm response strategies**—Assuming that some form of qualitative or quantitative analysis was performed, the project risks will have been prioritized either through a probability and impact matrix and/or tornado diagrams, as described earlier in this book. Beginning with the highest priority risks, a brainstorming session should be initiated to identify as many potential strategies as possible for each risk. Like any good brainstorming process, the emphasis should be on the quantity of ideas and criticism should be avoided. The most effective way to arrive at the best ideas is to have a lot of ideas. Depending on whether the risk is a threat or an opportunity, consider ways of implementing the various basic strategies identified in the following section of this chapter.

- **Evaluate response strategies**—Once a large number of potential responses have been identified for each risk, the focus should shift to evaluation. Each potential risk response should be discussed and evaluated by the group. The response strategies can be rated numerically (i.e., on a 1 to 10 scale where "1" is poor and a "10" is excellent) or simply be given a "yes" or a "no."

■ **Develop response strategies**—Those response strategies that are determined to have merit should be developed in detail. A narrative should be prepared discussing the details of the action plan and identifying responsibilities, timelines, and resources required to execute them. If a quantitative "postresponse" model is to be developed, the probabilities and impacts of the risk as modified by the response strategy should also be determined.

Risk Response Strategies

The actions available to address risks are based on the following risk response strategies to deal with threats:

■ **Avoidance**—The surest way to deal with a risk is to avoid it completely. There are a number of different ways to do this. One way is to modify the project scope.

  For example, assume that a particular retaining wall possesses a cost risk related to the geological conditions. If the retaining wall is eliminated, the risk can be avoided completely. However, the project cost may need to be increased in order to acquire additional real estate to allow for the replacement of the wall with an embankment. The question then is: Will the cost to avoid this risk be less than its expected impact? If the answer is yes, then this may be a good strategy to adopt. Many risks identified early on in a project's lifecycle can be avoided once additional information is developed.

■ **Transference**—Transferring a risk is a euphemism for "passing the buck." In other words, the risk can be passed on to another party, perhaps one that is more adept at dealing with a specific risk. Generally speaking, there is usually a price to be paid to do this. It is very common to pass on some risks to the contractor. The success of this strategy largely depends on the contractor's ability to assume and reduce the risk.

  For example, on a subway project one of the authors was involved with, the owner determined it would furnish the tunnel boring machine (TBM) equipment to the contractor. The risk management team felt that there was a great deal of risk associated with this approach as the contractor could blame any productivity problems on the owner-supplied TBM and file a construction claim. One strategy to deal with this risk would be to transfer it to the contractor by requiring him to furnish his own TBM. Of course, this risk transference will come at a price, but the risk management team's analysis indicated that the cost to do so was less than the expected impact of not doing so.

■ **Mitigation**—Risk mitigation is a strategy that does not prevent a risk, but rather reduces its probability and/or the severity of its impact. The appropriateness of risk mitigation is often related to the time in the project's lifespan when it is considered. Often it is easier to mitigate for risks early on and more costly to do so later in the project's lifecycle.

  For example, assume that a roadway project will require an extended period of heavy construction within 10 feet of several residences. If nothing is done to deal with this risk, it is likely that the affected residents will file a lawsuit, increase project costs, and more significantly, delay construction indefinitely. A mitigation response strategy for this risk would be to begin negotiations with the residents to temporarily relocate them for a period of time, thereby eliminating the chance for lengthy project delays. This particular mitigation strategy would definitely increase project costs, however, it mitigates for the risk by reducing its severity, especially in terms of schedule impacts.

- ■ **Acceptance**—The last strategy is to simply accept the risk. This is a viable strategy and may be appropriate for risks that are very small, very unlikely, or very difficult to respond to in using one of the previously mentioned strategies. It should be noted that this strategy assumes ''active'' acceptance, meaning that the appropriate risk reserves must be set aside to accommodate the risk's occurrence. In contrast, ''passive'' acceptance is simply to ignore the risk and hope it goes away—this approach is to choose not to manage the risk at all.

Often it is worth evaluating multiple strategies in dealing with risks, especially risks having a high expected impact. More often than not, the appropriate response is fairly self-evident. For larger projects, it is worth conducting a more comprehensive approach to developing risk response strategies by holding a value methodology (VM) workshop. In the authors' experience, a combination of risk analysis and VM has provided a very effective means of reducing project risk. There are creative solutions to dealing with risks; however, time must be devoted to doing this.

We must not forget about the risks that present opportunities for it is equally important to maximize these, just as we want to minimize threats. The following is a list of risk response strategies that apply to opportunities:

- ■ **Exploitation**—Opportunities possessing very strong potential benefits should be actively exploited. This is done by enhancing the probability that the opportunity will happen, or better yet, ensuring that it will happen. Often adopting this strategy will require some investment of project time and money to achieve, but if the return on investment is there, it will probably be worth it.

    For example, assume that an office building project has the opportunity to receive additional funding if it meets certain energy efficiency requirements. This opportunity could be exploited by making improvements to the building's HVAC and insulation systems. The risk management team should analyze the costs necessary to meet the requirements versus the additional funding it can receive. If the return on investment is there, we can enhance our chances of getting the funding by spending the additional funds on the improvements.

- ■ **Share**—Sometimes an opportunity can be capitalized on if we share the benefits with others. This speaks to the old cliché of creating a ''win-win'' situation. Most projects have many stakeholders with different objectives in mind. Often a little collaboration can go a long way in maximizing opportunities.

    An example of how to utilize the share strategy is through a value engineering incentive clause. Basically, this contract clause establishes a profit sharing mechanism between an owner and a contractor whereby the contractor is encouraged to develop cost saving modifications to the design. Cost savings are typically split, sometimes with the owner receiving the smaller share. The U.S. Army Corps of Engineers has been using this strategy for decades, resulting in millions of dollars in cost savings.

- ■ Enhancement—This strategy seeks to increase the probability of the opportunity of occurring and/or the degree of the resulting benefits. Enhancement is not always a sure thing, but often it can prove to be a worthwhile approach.

    For example, assume an opportunity is identified during risk analysis that indicates that there is a chance that the type of environmental document that is required could be changed for a major highway project. If the type of review can be reduced from an environmental assessment (EA) to a negative declaration (ND), the schedule could be accelerated by three months. This opportunity could be further enhanced if impacts were avoided to a certain area on the project. This could require a modification to the project scope or perhaps additional analysis. In any event, the chances of this opportunity occurring can be enhanced if specific actions are taken to do so.

Documenting Risk Response Strategies

As mentioned previously, a risk response strategy should be thoroughly developed and documented. Figure 5.3 provides an example of a risk response strategy for a highway project. It includes a description of the initial risk; the preresponse and postresponse probabilities and impacts; describes the initial risk response strategies; documents specific action plans describing how the strategies will be implemented; identifies the risk owner; establishes timelines for the action plans, meetings, reviews, and/or critical decision milestones; and identifies implementation costs and/or schedule impacts to implement the strategy.

Updating the Risk Register and Model

The risk register should be updated once the various risk response strategies have been identified. If the RBES is being utilized as the project's primary risk register, then it can be updated directly. Similarly, if a postresponse scenario is to be modeled, the adjusted probabilities and impacts should be identified as appropriate by the team. Examples of this process are provided in Chapter 6.

VALUE ENGINEERING AND RISK MANAGEMENT

In discussing risk response planning, there are other approaches that can and should be considered to enhance the quality and efficacy of identifying and developing risk response strategies. One such approach is known as value methodology (VM).

VM has existed under several different names over the years, such as value engineering (VE), value analysis (VA), and value management. There are no essential differences between these designations and they are, for all practical purposes, interchangeable. The term *value engineering* has been traditionally used whenever the value methodology is applied to industrial

Risk Response Strategies

Segment 1 - Risk Response Action Plans

VMS

Risk Description

Risk ID	Risk Name	Type	Description	Probability	Cost L	Cost M	Cost H	Cost EV	Schedule L	Schedule M	Schedule H	Schedule EV
10.2a	401 WQC - Stormwater	Threat	Stormwater mgt. Plan showing adequate TSS removal is needed prior to DNR issuing WQC. No WQC means you do not have a valid 404. Corridor analysis of stormwater management is behind schedule. This could potentially delay all projects with the primary concern being the earlier LETs.	50%	$0.00	$0.00	$0.00	$0.00	1.00	3.00	24.00	3.08

Risk Response Action Plan

Risk Response Strategies	Probability	Cost L	Cost M	Cost H	Cost EV	Schedule L	Schedule M	Schedule H	Schedule EV
1.Engage in early coordination by getting the final stormwater management plan in place ASAP 2.Re-evaluate current stormwater management based on current scope 3.Identify a US41 consultant lead to gain final approval of consultant contracts 4.Establish a drop-dead schedule date and manage to the schedule to ensure on-time delivery	25%	$0.00	$0.00	$0.00	$0.00	1.00	3.00	24.00	1.54

Action Plan Description(s)	Risk Owner	Risk Review Milestone / Frequency
1, 2. Early coordination = Consultant designer (Donohue) with close direction from Scott Eble along with the BC supervisor, 41 project managers, Tom Kobus (WisDOT NE Region stormwater) need to pick up where Hey and Associates left off and finalize/update the stormwater management plan to address current design concepts and staging. During recent coordination meetings with DNR it was made clear that the DNR is not satisfied with current design/analysis of stormwater management and will not be willing to give final WQC until we give him more information. To address the DNR's concern we should start by holding bi-weekly to weekly meetings with Donohue and other consultant stormwater pond designers to accelerate coordination and accelerate the completion of the final stormwater management plan. Then meetings should be held every month with the DNR to update them on their status and get input on the design. Scott Ebel as the assistant PM assigned to this task should take the lead in managing this issue and setting up and coordinating the agenda for these meetings. Courtney Chlopek should also be at these meeting to take notes and send out meeting minutes. 3. Identify US 41 lead to gain final approval of consultant contracts. 4. Scott Ebel schedule meeting as soon as Donohue contract is signed to setup schedule and set of milestones. Scott will then manage and update the schedule and milestones as the re-occurring meeting happen.	Lead - Scott Ebel (WisDOT)	Every other week review is needed. The drop-dead date for this is June 1, 2010, for segments 1 and 2. Note: Segments 1 and 2 must be delivered together.
	Base Cost Impacts	
	No cost impacts.	
	Base Schedule Impacts	
	No schedule impacts.	

Figure 5.3 Example Risk Response Strategy Documentation

design or to the construction industry; the term *value analysis* for concept planning or process applications; and the term *value management* for administration or management applications. *Value methodology* is the term most commonly used today and refers to the comprehensive body of knowledge related to improving value regardless of the area of application. Value methodology is formally defined as:

> A systematic process used by a multidisciplinary team to improve the value of projects through the analysis of functions.[2]

Value methodology is an organized process that has been effectively used within a wide range of private enterprises and public entities to achieve their continuous improvement goals, and in government agencies to better manage their limited budgets. The success of VM is due to its capacity to identify opportunities to remove unnecessary costs from projects, products, and services while assuring that performance, and other critical factors, meet or exceed the customers' expectations.

The improvements are the result of recommendations made by multidiscipline teams under the guidance of a skilled facilitator, commonly referred to as a value specialist. The multidiscipline teams can comprise those who were involved in the design and development of the project, technical experts who were not involved with the project, or a combination of the two. There are two essential elements that set VM apart from other techniques, methodologies, and processes:

1. The application of the unique method of function analysis and its relation to cost and performance
2. The organization of the concepts and techniques into a specific job plan

These factors differentiate VM from other analytical or problem-solving methodologies.

VM is often confused with cost reduction; however, cost reduction and VM are distinctly different. Cost reduction activities are component-oriented. This often involves the act of "cheapening" the item—that is, reducing cost at the expense of performance.

Value methodology, conversely, is concerned with how things function rather than what they are. This function-driven mindset demands a radical transformation in our perception; in the way we approach challenges, both old and new. This functional way of thinking is, by its nature, predisposed to lead us to innovative solutions by opening our eyes and deepening our understanding of how things work. This concept of function is the very essence of value methodology.

Typically, VE studies are conducted on a project at one or more points during its design development. They involve the use of independent, multidiscipline teams that are led by a skilled facilitator following a specific job plan. Traditionally, these efforts have been independent, preplanned events that do not involve risk analysis.

Based on years of experience, however, the authors strongly recommend that risk assessment and VE be integrated. The two processes are a natural fit—risk assessment identifies and characterizes the problems while VE generates and develops the solutions.

Integrating Value Engineering and Risk Response Planning

The discipline of risk management has traditionally focused on the identification and quantification of risk. This focus on problems, while it has indeed proven to be effective in improving decisions that involve uncertainty, often misses the mark with respect to identifying appropriate responses in an innovative way that maximizes project value. Value improvement relative to the management of risk requires that attention be given to project functions. The integration of function analysis into risk management provides a powerful means to do this.

Similar to most risk workshops, an integrated VE and risk assessment workshop utilizes a multidiscipline team composed of subject matter experts (SMEs) representing various areas of knowledge relevant to the project. In addition, it requires a facilitator who is also fluent with function analysis as well as other VE techniques. This may be the same person as the risk manager or a co-facilitator who will work in conjunction with the risk lead. It is important to note that a skilled facilitator with the necessary qualifications is important, as they will ultimately be able to best drive the process and achieve the desired outcome.

Although VE techniques can be applied throughout the risk assessment process,[3] this book will focus on its use during risk response planning. It is recommended that practitioners consider the following approach to integrate function analysis with the traditional risk response planning process. The following steps comprise this integrated phase:

- **Risk Object Identification**—The affected area of impact is best described by the *risk object*. The risk object is the area affected by the risk in relation to the activity or project function being impacted. It is effectively also the elemental nature of the risk that can be managed. The object of risk for each individual risk is identified and utilized as the management target for idea development for risk response strategies. The risk object is typically the noun from the impacted project or system function, which is comprised of a two-word abridgement of a verb-noun combination.

  This activity is essential to the process as the articulation of risk event descriptions into simple, concise statements of risk comprised of no more than a few words helps to focus the team on the problem. It has been the authors' experience that participants tend to lack focus on what the actual object of risk is—while there may be a lengthy, detailed description of the event, it is still sometimes difficult to distill a concisely described risk event. This lack of clarity inhibits the ability of the team to identify solutions.

- **Brainstorming of Risk Response Strategies by Function**—The brainstorming of risk response strategies by function is a three-step process. The first step is to establish the risk object, which becomes the target element that can effectively be managed, and it is also the element to which a risk response will provide the most direct buffering of risk impacts. Second, brainstorming of risk response strategies are developed by identifying the risk response function. The risk response function is a verb-noun combination that describes the risk response strategy to be employed. Third, a brief description of each idea is provided for each response strategy. Throughout the process of brainstorming, each high and moderate priority risk should receive attention. Also, the brainstorming process includes identification of specific strategies in the form of the function/verb that are possible to use, depending on whether the risk is a *threat* or an *opportunity*. For threats, the following function verbs are possible:

- Accept
- Avoid
- Mitigate
- Transfer

  For opportunities, the following function verbs are possible:

- Enhance
- Exploit
- Share

  Figure 5.4 illustrates the process of transforming a risk description into a function statement. The process is similar for both threats and opportunities; however, the verb selection varies, as described above. The activity provides a simple, concise problem statement that will allow participants to visualize many potential solutions rather than just one or two.

- **Evaluation of Risk Response Strategies**—The evaluation of risk response strategies are brainstormed and evaluated to determine which responses provide the most relative value to either minimizing threats or maximizing opportunities. Each response strategy should be qualitatively evaluated in relation to this criterion. For example, each response strategy can be given a green check mark, a yellow exclamation mark, or a red "X." The response strategies that have green checks become the risk response strategies that are developed in further detail. The yellow exclamation marks become fall-back strategies that could be put into place as efforts to manage the risk if the preliminary strategies are not working as

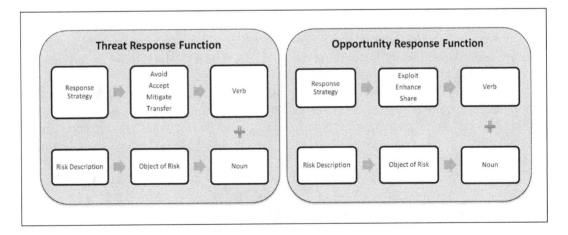

Figure 5.4   Converting Response Strategies and Risk Objects into Function Statements

effectively as anticipated. The yellow exclamation marks also have the possibility of being developed as additional risk response strategies. The red Xs are deemed to be invalid or ineffective risk response strategies to utilize in the context of the project. Keeping the evaluation simple is best in this case so that the time in the workshop can be most effectively utilized. Figure 5.5 provides an example worksheet detailing the results of this process.

- ■ **Development of Action Plans for Risk Response Strategies**—The final development of the risk action plans involves a combination of several elements. This includes the assignment of the risk to key individuals or groups that are deemed to be best equipped to manage or deal with the risk by the risk assessment workshop team. Development of action plans also includes providing more detail around the risk response strategy selected in the form of developing specific actionable steps that can be followed in order to best manage the risk.

This is just a partial example of how VE can be integrated into the RBE process. Additional information on VM/VE techniques is available through the book *Value Optimization for Project and Performance Management*.[4]

Disaster Contingency Planning

Disaster contingency planning is a deterministic approach to risk response planning that assumes a threat will occur and seeks to identify contingency and/or recovery responses to deal with the aftermath. Contingency planning is an excellent approach to manage extreme events that could have significant and severe consequences for a project. These project-related Black Swans should be addressed in some fashion and should not simply be ignored as "uncontrollable." While there indeed may be no control over such risks, there are always ways to help reduce the severity of their effects should they occur. This approach to risk response planning should be considered as a complementary form of planning that can be utilized as needed and appropriate.

Contingency planning is about "hoping for the best, but planning for the worst." Contingency plans must be actionable—they cannot be so outlandish that they would never be implemented. They are best aligned with the "accept" strategy but are more fatalistic in that they assume that the risks will occur rather than treating them as uncertain events.

Contingency planning played a major role in U.S. foreign, domestic, and defense planning during the Cold War era. A great deal of time, money, and effort was expended in planning for nuclear war. The assumption in Washington was that it *would* happen—the challenge was to implement strategies to find ways for citizens and their governments to survive and rebuild in the aftermath of a nuclear war.

The Deepwater Horizon catastrophe discussed in Chapter 1 is an excellent example of the result of a lack of contingency planning. Had contingency plans been identified for such a disaster, it is likely that the size and extent of the spill would have been much smaller. To be sure, there are costs involved to implement contingency plans, however, in the case of Deepwater Horizon, it is likely that had either British Petroleum, or the oil industry as a whole, invested in

Risk Response Strategies
Team Brainstorming

VMS

Risk ID	Risk Name	Type	Description	Object of Risk	Probability	EV Cost	EV Schedule	Function Verb	Function Noun	Ideas	Rating
23.1	TransCanada at Main Ave/CTH G	Threat	A critical TransCanada pipeline crosses Main Ave/CTH G just west of the southbound ramp terminal, then follows the west side of the SB off-ramp and crosses US 41. This pipeline may be in conflict with the proposed storm sewer system, pond excavation, and street lights. TransCanada typically requests a 25-foot-wide clear zone between their pipelines and any structures, such as manholes, inlets, and light poles. Whereas this line is primarily within the right-of-way, it's compensable and TransCanada is slow and expensive to move.	Pipeline Conflicts	75%	$1.31	1.25	Mitigate	Pipeline Conflicts	Aggressively and clearly define all conflicts with the TransCanada pipeline	◎
								Mitigate	Pipeline Conflicts	Early coordination with TransCanada to identify pipe location	◎
								Mitigate	Pipeline Conflicts	Investigate legal coordination issues early on with TransCanada to establish who was there first and what the compensation levels were	◎
								Avoid	Pipeline Conflicts	Redesign intersection to eliminate any conflicts with the pipeline	◎
								Avoid	Pipeline Conflicts	Move to a tight diamond configuration to shift profile (assuming both Ashland Ave. structures are being replaced)	◎
23.2	Expedited Utility Coordination	Threat	The risks associated with violating TRANS 220 timelines or the need to fairly compensate utilities may be outweighed by the need to minimize construction delays. Thus DOT would take on liability for construction delays due to unresolved utility conflicts.	Utility Liability	95%	$1.20	2.53	Avoid	Utility Liability	Follow TRANS 220 process.	◎
								Accept	Utility Liability	Violate TRANS 220 process for the trade-off of getting projects built earlier and allocate necessary funds to budget	◎
								Mitigate	Utility Liability	Engage in extra coordination with utilities	◎
								Mitigate	Utility Liability	NCG monthly meetings with Beecher Hoppe for coordination	◎
								Transfer	Utility Liability	Shift responsibility to contractor and have him price risk in bid	◎
24	Re-estimating Quantities	Opportunity	There is the opportunity to realize cost savings resulting from re-estimation of quantities	Cost Savings	75%	$6.88	0.00	Exploit	Cost Savings	Get consultant under contract to provide updated estimate to validate cost savings	◎
								Exploit	Cost Savings	Review upcoming bids to "learn" what unit prices are looking to be and what the contractor bidding environment is possibly looking to be	◎

Figure 5.5 Example of a Risk Response Brainstorming and Evaluation Session Using Function Statements

modest contingency and recovery plans, the cost would have been a mere fraction of the $20 billion placed in escrow to settle claims.

The authors were involved in a risk assessment for a very large project (over $1 billion). An elaborate quantitative risk analysis was performed for the project, which was very comprehensive in terms of the number and types of risks identified and assessed. Despite all of the effort put into the risk elicitation, modeling, and analysis, the project failed, and for reasons that lay completely outside of the model. The risk that doomed the project was the stock market crash of 2008, which caused a loss in the majority of funding, resulting in the cancelation of the project. This risk, obviously, could not have, nor should it have been, modeled. However, had some level of basic contingency planning been performed, a number of strategies could have been identified that could have salvaged and/or better utilized the project funds that had been allocated and spent.

RISK MONITORING AND CONTROL

Risk monitoring and control is concerned with the active management of risks following the previous five steps of the risk management process. The various activities included in this step include:

- Tracking risks on the risk register
- Identifying new risks and retiring old ones
- Adjusting risk response strategies or developing new ones
- Managing risk contingency reserves
- Monitoring the execution and effectiveness of risk response strategies

This step continues throughout the remaining life of the project and is absolutely essential. This is the part of the process where the emphasis is on "doing" rather than "discussing."

Monitoring Risk

The process of monitoring project risks is the first component of this phase of risk management. There are a number of fairly basic components of this process. These include:

- Identify who is the owner of the risk.
- Identify who is responsible for monitoring the risk. For the purposes of this text we will call this person the *task manager*. The task manager will be responsible for tracking the effectiveness of the plan, identifying any unintended consequences, and making suggestions on mid-course corrections to further mitigate the risk.
- Identify the nature and frequency which the task manager for each risk will report to the project manager (or whoever the designated risk manager is for the project).
- Develop a monitoring protocol with respect to the form of these updates. It is a good idea for each task manager to submit their updated risk information to the project manager

so he or she can update the master risk register. It is important to maintain a master document in order to ensure that all of the different updates coming from different sources are consolidated.

■ The project manager should distribute updates of the risk register to the risk management team as they develop, or establish a schedule for regular status updates (i.e., weekly, monthly, and so forth).

One of the most effective tools in monitoring risks and their responses is the development of a risk monitoring schedule. Figure 5.6 is an example of this. Essentially, each risk is added to a master schedule using scheduling software such as Primavera® or Microsoft Project®. The specific action plans identified in the risk response strategy documentation are noted and specific dates are targeted for processes, meetings, and key milestones. The vertical dark line indicates the last status update. The risk manager can use this tool to check with the various risk owners what the status is and hold them accountable for their responsibilities.

Another risk monitoring approach includes the development of a risk management information system, which is essentially a risk database. This approach is recommended for very large projects with many risks and/or many potential response strategies and action plans. The database structure is similar to that of a risk register; however, more detailed information can be maintained relative to schedule and status. Figure 5.7 shows a screenshot for a risk management information system developed for a highway project.

As risks are addressed during project design and delivery, the risk reserve fund (i.e., contingency) must also be managed as funds are allocated to implement specific risk response strategies and any risk-related cost or time impacts. This information should also be monitored as the project moves through its lifecycle and adjustments made as necessary.

Controlling Risk

Projects change as they move through their lifecycle. Sometimes these changes are radical in nature. It is therefore important to validate the risks on the risk register as the project evolves. For example, the project's scope could change significantly or the schedule could change unexpectedly. In these cases, it will be necessary for the risk management team to reevaluate the risks.

It may be necessary to hold more than one workshop on large, complex projects; these should be included in the schedule to ensure that they are not delayed until the very end. There are three project milestones where this makes good sense. These include:

■ **Concept development**—Conducting a risk workshop at the end of concept development will be useful in finalizing the initial design approach. Sometimes, large risks emerge that were not initially identified. Usually the large "project-killer" type risks are identified at this stage. In such cases it is not uncommon for this to result in significant design changes or even lead to a switch to a different design concept.

■ **Preliminary design**—A risk workshop should be planned sometime during the preliminary engineering or design phase. The timing of this will depend on when, or if, a risk

Task ID	Mitigation Task Description	Rem Dur	Current Start	Current Finish	Variance From Plan	Mitigation Status	Other Risk Owners
GBMSD							
C221-1.4	Stay w/ ROW between Memorial and 1500' North	0	08FEB10A	01APR10A	16		
C221-2.0	Set dates for 60% delivered for GBMSD sgmts	0	08FEB10A	01APR10A	16		
C221-4.0	Dvlp prelim bridge plans w/no conflicts w/subustr	0	08FEB10A	01APR10A	16		
C18.4-1.6	"ADO" Deliver Recorded Plat (2 ea)	20	01APR10	28APR10	0		
C18.4-1.8	"ADO" Sign/Deliver Utility Agreement	24	29APR10	01JUN10	0		
C18.4-1.10	"ADO" Response letter on Soils	21	02JUN10	30JUN10	0		
C022.1A	GBMSD	0		01JUL10	-49	Process Est.	
Unbundling							
C00.2-2.1	Explore issues w/unbundling lighting	60	08FEB10A	23JUN10	-43		
C00.2-1.0	Transfer funds into appropriate contract&year	60	08FEB10A	23JUN10	-43		
C00.2-2.2	Look at RAB staging plan, for unbundling option	60	24JUN10	15SEP10	-43		
C00.2-2.3	Add nonunbundled lighting to orig contract	60	16SEP10	08DEC10	-43		
C00.2-2.4	Req contractors coord windows for install	60	16SEP10	08DEC10	-43		
C000.2	Unbundling	0		08DEC10	-222	Pending	

Figure 5.6 Example of a Risk Monitoring Schedule

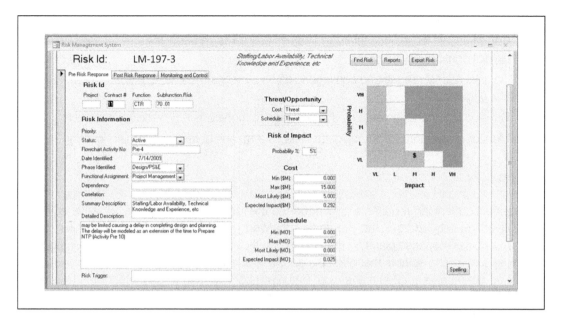

Figure 5.7   Example of a Risk Management Information System

workshop was held at the concept design phase. A risk workshop held at this phase will generally identify numerous risks related to design development issues and includes things like schedule delays related to technical reviews, real estate acquisition, and uncertainties related to the design.

- **Final Design**—It may be wise to conduct a risk workshop sometime during the final design phase. Such workshops are generally geared at focusing on construction- and contract-related risks.

Interim risk reviews should be scheduled on a regular basis between these major project milestones and during construction as necessary. Updates can be made as necessary to existing risks; modifications to ongoing risk response strategies can be implemented; identification of new risks and the development of new response strategies to deal with them can be exercised.

The National Cooperative Highway Research Program (NCHRP) identified a number of keys to success for cost estimating and risk management.[5] These include:

1. Complete every step in the estimation process during all phases of project development.
2. Document estimate basis, assumptions, and backup calculations thoroughly.
3. Identify project risks and uncertainties early, and use these explicitly identified risks to establish appropriate contingencies.
4. Anticipate external cost influences and incorporate them into the estimate.
5. Perform estimate reviews to confirm that the estimate is accurate and fully reflects project scope.

6. Employ all steps in the risk management process.

7. Communicate cost uncertainty in project estimates through the use of ranges and/or explicit contingency amounts.

8. Tie risks to cost ranges and contingencies as a means of explaining cost uncertainty to all stakeholders.

9. Develop risk management plans and assign responsibility for resolving each risk.

10. Monitor project threats and opportunities as a means of resolving project contingency.

SUMMARY

Risk response planning requires a concerted team effort in order to identify effective solutions that will adequately address the problems identified during risk assessment. Similar to risk assessment, risk response planning requires that sufficient time and resources be allocated. It is the experience of the authors that often the activities of risk response planning, monitoring, and control do not receive adequate attention. Returning back to the bridge metaphor in Chapter 1, risk assessment carries us only halfway across the river. Risk response planning, monitoring, and control are necessary to complete our journey.

The application of value methodology is a particularly effective vehicle for identifying and developing broad strategies and specific action plans to deal with risks.

One of the key tenants in effectively managing risks is assigning specific responsibilities to people and then holding them accountable. There is an old saying that states, "If it's everybody's job, it's nobody's responsibility." This is especially true when it comes to effective risk management. Risks are abstract and ephemeral by nature and many people have a hard time taking them seriously because they either don't believe they can happen on their project or, if they do, that they can be easily addressed. As was discussed in the previous chapter, there are numerous cognitive biases that further exacerbate this problem.

Effective risk management requires the same diligence as managing scope, schedule, and cost as unchecked risks are likely to have unexpected impacts to each of these areas.

ENDNOTES

1. A. Tversky and Daniel Kahneman, "Choices, Values and Frames." *American Psychologist* 39 (1984), 341–350.

2. SAVE International Value Standard (2007).

3. R. Stewart and G. Brink, "Function Driven Risk Management." *Value World*, Summer 2010, SAVE International.

4. R. Stewart, *Value Optimization for Project and Performance Management*. Hoboken, NJ: John Wiley and Sons, 2010.

5. National Cooperative Highway Research Program, Report 658, *Guidebook on Risk Analysis Tools and Management Practices to Control Transportation Project Costs*, 2010. Washington, DC: Transportation Research Board, 2010.

CHAPTER 6

RISK-BASED ESTIMATE
SELF-MODELING SPREADSHEET

OVERVIEW

The risk-based estimate is a process of cost risk analysis that requires the employment of the Monte Carlo method as a way of creating data for statistical analysis. The Monte Carlo method (MCM) is a well-recognized statistical approach of evaluating and analyzing complex and cyclic events. Just Google it and you will find thousands of sites, one better than the next, that relate to the Monte Carlo method or Monte Carlo simulation, or Monte Carlo casino. You may find the story of how the method was invented and then developed; you may find different software programs that facilitate the method, and you may find our risk-based estimate self-modeling spreadsheet (RBES). You may even find a way to improve your odds at a Las Vegas casino.

What makes RBES stand out among other cost and schedule estimate software is its simplicity and flexibility to mold the model to a project's particularities. RBES uses only Excel features and an Excel user is able to develop a case model and run cost and schedule risk analyses. Once the model is created, RBES allows for continuous execution of what-if scenarios to better understand and compare a risk's impact.

The ability to have a project manager run what-if scenarios has proven many times to be an extremely powerful tool. What the authors have found is that it encourages buy-in on the entire risk-based estimating and management process by project managers, because they can see first-hand what effects risks have on their projects. Project managers get the feeling that the results are not just something out of a black box; they can actually understand the impacts of risk, which leads to better monitoring, control, and management.[1]

While RBES was developed to address the needs of the transportation industry, it is now being adopted by project managers in different fields (large hospitals, ferry terminals, industrial plants, safety analysis, and enterprise risk management). RBES can be used in any direct effort to analyze the effect of risk and uncertainty on a project's cost and schedule. Additionally, RBES is a tool that facilitates ongoing risk management practice by allowing the user to compute and display results of premitigated and postmitigated risk analysis. Ultimately, RBES hides the science of risk analysis and becomes an indispensable tool for cost and schedule estimation.

The RBES has an algorithm based on the flowchart diagram presented in Figure 3.42. If project conditions require change of the flowchart diagram then RBES can adapt and adjust its algorithm. This change usually requires an experienced modeler's work but once it is done the model may be used continually by regular users.

The following information relates to RBES as defined by the flowchart diagram presented in Figure 3.42. We go into details sufficient to show the RBES's functions and examine different uses, but not enough to serve as the RBES's instruction manual.

Figures 6.1 and 6.2 show how RBES is organized. Figure 6.1 presents the model through its inputs and outputs and Figure 6.2 presents the model in schematic form, focusing on the model's components and their interactions. The model has two major branches: (1) premitigated analysis and (2) postmitigated analysis that can run in sequence. Premitigated analysis uses the data provided in its original form: It represents data before project team members have had a chance of planning for risk response actions. The postmitigated analysis, therefore, always follows premitigated analysis. Results are shown either through overlay graphs or side-by-side tables.

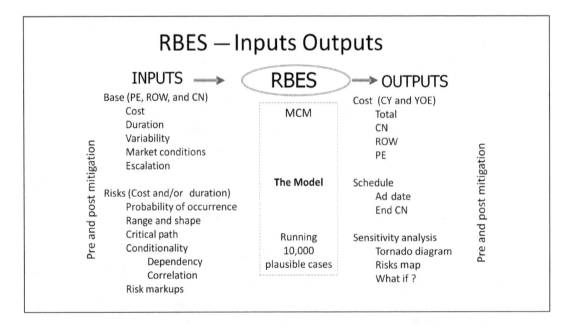

Figure 6.1   Risk-Based Estimate Self-Modeling Spreadsheet Inputs and Outputs

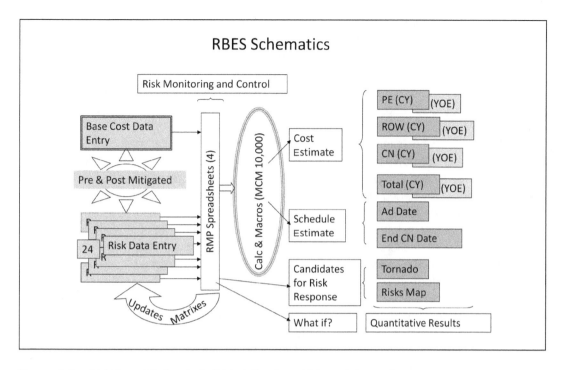

Figure 6.2   Risk-Based Estimate Self-Modeling Spreadsheet Schematics

A project's cost and schedule data can be separated into two major components: (1) base cost and schedule and (2) cost and schedule risks, and they apply to each of the three phases of the flowchart: (1) preliminary engineering (PE), right of way (ROW) (which is the right to use land), and (3) construction (CN).

Base cost comprises: (1) base variability, (2) market conditions, and (3) inflation rates, which are captured on the base cost tab. The base cost and schedule data are critical for any cost risk analysis and any error in their numbers is translated linearly in the model's final results.[2]

A cost and/or schedule risk has the following attributes: (1) probability of occurrence, (2) impact distribution, (3) dependency among other risks, (4) correlation with other risks, (5) critical path for schedule risks, (6) risk markups (captured on the base tab), and (7) the activity for which it is applied (these are captured on the risk sheets). The risk markups are factors applied to construction risks impact values in order to emulate the base cost estimate structure (i.e., preliminary engineering, mobilization, sales tax, construction engineering, and change order contingencies).

The model output has three major tiers: (1) cost, (2) schedule, and (3) sensitivity analysis. The output cost distribution is presented for a project's phases in two manners: (1) current year (time of estimate), and (2) year of expenditure (time when money is spent). The cost distribution is then presented in the form of graphs (mass diagram and cumulative distribution function) or tables (percentiles tables).

The schedule results present the advertisement date and end of construction date in forms of graphs and tables. The sensitivity analysis indicates candidates for mitigation in the form of a tornado diagram for cost and schedule and a project's risks map.

Figure 6.2 presents the structure of RBES from a different angle. The left side of Figure 6.2 presents the entry data for premitigated and postmitigated phases. The data entered feed the RMP spreadsheet (RMP and RMPSuppl capture risks defined in the premitigated phase and RMPM and RMPSupplM capture the postmitigated phase).

The RMP spreadsheets include risks' assessment data in quantitative and qualitative form and information about how risks may be treated (response plan). When these spreadsheets are filled in correctly and completely, they constitute a powerful communication tool that helps the process of risk management. There are project managers who carry in their pocket this 11-by-17 inches spreadsheet print and do not miss any opportunity to refer to it.

The model's engine is in the Calc spreadsheet, which, together with the workbook macros, orchestrates the computations and results presentation. MCM runs 10,000 iterations, which is sufficient for these kinds of calculations. The running time is less than one minute depending on your computer's performance.

The results presented by RBES basically are in forms of histograms, cumulative distribution functions, and tables. Each critical project phase may have its cost distribution in current year dollars or year of expenditure dollars. The advertisement date (Ad Date) and end of construction date (End CN Date) have their own estimated distribution. The candidates for risk response are depicted through two diagrams: (1) a tornado diagram for cost and for schedule, and (2) a risks map, which differentiates cost, schedule, threat, and opportunity.

The results as previously shown are computed for premitigated, and then when postmitigated data is available, the postmitigated results are placed side by side on a table and overlaid on graphs. This feature is wildly used when cost risk analysis is combined with a value engineering study.

MODEL ACCURACY

A fair question has been raised about the level of resolution that the RBES may provide. The concern is derived from the fact that RBES uses random numbers generated by Excel. Other specialized software programs use more sophisticated random number generators and more advanced sampling methods such as Latin-Hypercube. Specialized software programs offer the option of selection between MCM sampling and Latin-Hypercube sampling.

The sampling methods for MCM were developed at a time when computing speed was limited, and even for simple models the simulation time was significant. Latin-Hypercube sampling uses a technique known as stratified sampling without replacement[3] and it has the advantage of generating a set of samples that reflect the shape of a sampled distribution versus the pure random (Monte Carlo) samples, which is quicker. The general effect is the speeding of the process of simulation by requiring a small number of iterations.

The RBES random number generator works like MCM sampling and the program doesn't offer the Latin-Hypercube sampling option. This "limitation" has raised suspicions among users that RBES may not be able to provide enough precision to its results. As we have presented in the previous paragraphs, the method of sampling has nothing to do with model accuracy; it only affects the number of iterations needed and the simulation run time.

To clarify this issue the authors have run and compared the results of a simple Pert distribution of "$100 M ±20%." In other words, it was tested in a Pert distribution with a minimum value of $80 M, a maximum value of $120 M, and a most likely value of $100 M. The RBES and specialized software (@RISK for Excel) were used to run a simulation that comprises 10,000 iterations.

The results as presented in Table 6.1 enforce the statement that the random sampling does not affect the quality of the model results. Table 6.1 shows that the results produced by both sampling methods are in narrow proximity of each other. Only a few hundredths of a percent or even one-tenth of a percent apart will not likely make a difference for application to a cost risk

TABLE 6.1   Percentiles Results RBES and @RISK

RBES vs. @RISK	
RBES	@Risk
100 +20%	100 +20%

Statistics	RBES	@Risk	Delta
Min	80.28 $M	80.69 $M	0.50%
Max	119.07 $M	119.29 $M	0.18%
Median	100.03 $M	100.00 $M	-0.03%
10%	89.69 $M	89.86 $M	0.19%
20%	93.06 $M	93.06 $M	0.01%
30%	95.62 $M	95.59 $M	-0.03%
40%	97.91 $M	97.85 $M	-0.06%
50%	100.03 $M	100.00 $M	-0.03%
60%	102.18 $M	102.15 $M	-0.03%
70%	104.45 $M	104.41 $M	-0.04%
80%	106.98 $M	106.93 $M	-0.05%
90%	109.98 $M	110.13 $M	0.14%

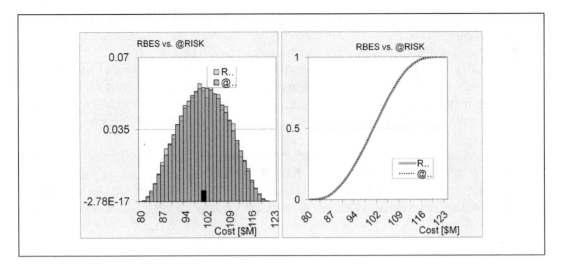

Figure 6.3   Histogram and CDF Results of RBES and @RISK

assessment. If the same model uses Latin-Hypercube sampling for two simulations, but with different seed numbers for each simulation the results will not be identical.

The proximity of these two distributions is clearly presented in Figure 6.3. While the histogram of both distributions may show some dissonance (the RBES histogram is rougher than the @RISK histogram), the cumulative distribution functions are identical. It is hard to see if the graph contains one or two curves.

RBES—COST AND SCHEDULE DATA INPUT AND OUTPUT

In Chapter 3 we discussed that RBE is a process that is based on two main components: (1) base cost and schedule review and (2) elicitation of risks' events that may affect project cost and/or schedule. RBES follows the process of RBE and clearly marks the base cost and schedule data and risks events through allocating a spreadsheet (tab) for base cost and schedule information and a spreadsheet for each risk event analyzed. RBES may use four types of distributions: (1) normal, (2) uniform, (3) Pert, and (4) triangular. These distributions may be assigned to the base cost and schedule and the risk's impact. The Pert distribution is the default distribution and it is the distribution that the free copy of the simplified model you may find on the Internet site.[4]

Because of close interdependency between base data and risks data, some of the information presented on the base cost and schedule spreadsheet may be used to complement risks in the process of simulation.

Base Cost and Schedule Data Input

Base cost and schedule is the most important component of any RBE. As Chapter 3 has presented, the base data is crucial to the accuracy of the entire analysis. Any error introduced in the assessment through base cost and schedule will linearly be represented on the final results with no way to identify it.

Base Data Entry Tab

The Excel spreadsheet that captures the project base cost and schedule is named "Base" and is presented in Table 6.2. The data is entered in shaded cells (in the electronic view, it is a yellow shade) and each cell is associated with a validation pop-up window that explains the cell's content. There are empty cells that may be used for custom design of the model to meet specific requirements. Some of the entry cells are quite self-explanatory and we will not spend time explaining them.

The adjacent cell to "Estimate Date" indicates the date of the time of the estimate. This date (in our example 10/10/08) is important because it represents the estimated unit costs at that time and the model uses this date as a reference (start) point for calculating the inflated values of each project activity.

The adjacent cell to "Last Review Date" indicates the last time when the PM reviewed the model without running a new simulation. It captures the discussions on risk response and monitoring and control. If during the process of monitoring and control the base cost is changed because of unit prices, then the "Estimate Date" needs to be updated and the model rerun.

The adjacent cell to "Target Ad date" shows the target advertisement date (01/01/09) when the project is scheduled to be ready for bids or letting. Next to it, the variability of the interval between "Target Ad Date" and "Estimate Date" is presented. The RBES allows only symmetrical distribution of base variability. For example, 5 percent entered in the cell adjacent to "Target Ad Date" indicates that the duration interval between "Target Ad Date" and "Estimate Date" has

TABLE 6.2   Base Cost and Schedule Data Entry Sheet

Project Title			Highway to Heaven	Value	Variability	Risk Markups		Inflation tables built-in	
Estimate Date	10/10/08		Target AD date	01/01/09	5%	Mob	10.0%	A/B/A Duration	2Mo
Project PIN #			Estimated CN Duration	12.0Mo	10%	Tax	9.8%	Annual inflation rates	YOE
Last Review Date	12/12/08		Estimated PE Cost	20.00 $M	5%	CE	12.0%	PE	4%
Project Manager			Estimated ROW Cost	38.00 $M	5%	PE	10.0%	ROW	6%
WSDOT accepts no responsibility for its use			Estimated CN Cost	200.00 $M	10%	C.O.C	4.0%	CN	3%
Inflation Points			Base CN Cost Market Conditions			Distribution Type		Base ROW Cost Market Conditions	
Define inflation point of the activity cost. For example 50% means that the inflation point for that activity is the mid-point activity. 50% is the default value. If it is decided that the inflation point is at three quarters of respective activity enter 0.75.		Better than planned	10%	10%		PE Duration	U	Better than planned	10%	5%
		Worse than planned	15%	20%		CN Duration	U	Worse than planned	20%	10%
			Probability	Impact		PE Cost	P		Probability	Impact
Preconstruction activities (ROW and PE)	0.5					ROW Cost	T			
Construction	0.5					CN Cost	T			

N=normal distribution; U=uniform disgtribution; P=PERT or Beta3 distribution; T=triangular distribution.

a range of ±5 percent. In the case presented in Table 6.2, the preconstruction duration is 78.8 days and the model uses values between 12/27/2008 and 1/5/2009.

The adjacent cell to "Estimated CN Duration" shows the base duration of the entire construction activity (12 months) and 10 percent located in the next cell indicates that the construction activity may range from 10.8 months [12 × (1 − 0.1)] to 13.2 months [12 × (1 + 0.1)].

The next three cells capture: (1) estimated base cost for preliminary engineering (PE) cost (the estimated cost for making the project ready for construction ($20 M), (2) estimated base cost for the right of use of land (ROW) (includes all costs related to acquisition or agreements including cost to the lawyers [$38 M]), and (3) estimated base cost for construction ($200 M). Next to these values there are cells that indicate the variability of related values: (1) 5 percent for estimated PE cost, (2) 5 percent for estimated ROW costs, and (3) 10 percent for estimated CN costs.

That means that the base PE estimated cost ranges from $19 M to $21 M, the base ROW estimated cost ranges from $36.1 M to $39.9 M, and the base CN estimated cost ranges from $180 M to $220 M.

The section in the upper-right corner presents information about inflation rates. For example, the project may have assigned: (1) 4 percent annual inflation rate for PE cost, (2) 6 percent annual inflation rate for ROW costs, and (3) 3 percent annual inflation rate for CN.

RBES offers users the ability to select the point in time when inflation is performed. The lower-left corner allows for entering information about the position in time of the point of inflation. Figure 6.4 shows how the inflation point is defined for preconstruction and construction. For example, the schedule values (0.5) entered in Table 6.2 define the inflation point of preconstruction activities at 01/01/2009 (which is the same with the target ad date) and inflation point of construction activity at 07/03/2010 (which is six months after the target ad date). The

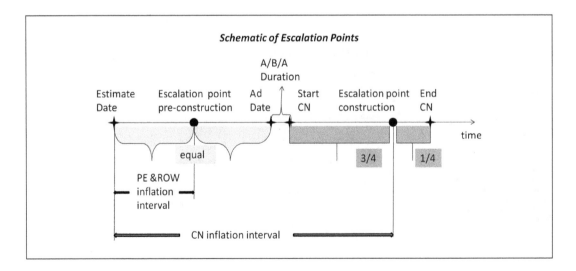

Figure 6.4   Schematic of Inflation Points

values presented are for base estimates only and any identified schedule risks will change the dates of each iteration to capture the risks' schedule impact.

Risks Markups

The risks markups section affects all risks that apply to the construction activity. The numbers represent percentage values used for defining the deterministic cost estimate. The deterministic value of the construction base cost estimate includes elements such as the cost of contractor's mobilization (Mob), sales tax (Tax), construction engineering, which is the direct cost to the owner for administering the construction (CE), and change order contingency (COC), which represents a small contingency used for handling small change orders. When construction risks occur they may introduce a new bid item or may add cost to an existing bid item. The additional cost that the construction risk introduces is elicited as its face value (without markups). In order to capture the construction risk's real effect on construction cost, the elicited risk values need to be adjusted by the same factors as any other bid item.

The risk markups section shows the markups' value used when the construction cost is calculated and these values will affect the construction risks when they occur. Table 6.3 presents the summary of project deterministic base cost estimate and how markups' factors are applied on our example.

TABLE 6.3   Project's Markups

Estimate Summary		
Construction Costs plus Mobilization		133 $M
Design Allowances	30.0%	40 $M
Change Order Contigency (C.O.C.)	4.0%	7 $M
Sales Tax	9.0%	16 $M
Construction Engineering (CE)	8.0%	16 $M
Construction Total		212 $M
Right of Land Use (ROW)		31 $M
Preliminary Engineering (PE)	9.5%	20 $M
Total Project Costs		263 $M

Each organization has its own recommendations on markup values and has its own scheme of how the markups are to be applied. It is important to have consistency on how the markups are applied on the deterministic base cost estimate and how they are applied during the risk-based estimating process.

Risks Data Entry

Overview

The RBES allows up to 24 risk events to be entered and analyzed for premitigated and/or postmitigated scenarios. Risks are events that have their own identities; sometimes they may have a complex identity (see SMART, Table 3.6), which needs to be captured and recorded. In order to allow users to focus on every risk without staring at the data of other risks, the authors have created 24 risk sheets (tabs) that permit recording of the individual risk's description, quantitative impact, matrix qualitative display, and comments about any pertinent data.

Figure 6.2 indicates that the risks data contained on each risk sheet is placed on the RMP and RMPSuppl tabs for the premitigated scenario and RMPM and RMPSupplM tabs for the postmitigated scenario where the risk matrix qualitative display is updated and the risks' conditionality is enforced. We will call RMP, RMPSuppl, RMPM, and RMPMSupplM the "big four" tabs.

Risks Assessment

The 24 risk sheets contain risk assessment data that users put into the model including suggestions about how the risk may be optimized (minimized or terminated, if it is a threat and maximized if it is opportunity). The big four tabs comprise the risk data and risk response plan including monitoring and control.

The big four tabs have two distinct areas: (1) risk assessment and (2) risk management (response planning and monitoring and control). Bringing these two major areas of risk management together on the same piece of paper allows users a quick comprehensive view of project risks. The risk assessment section comprises three areas: (1) risk identification, (2) quantitative analysis, and (3) qualitative display of the event's expected value.

The risk management section includes two areas: (1) risk response planning and (2) risk monitoring and control. Later in the chapter we present more information about the management sections of the big four tabs.

Risk Data Entry Tab

Table 6.4 presents the upper section of the risk data entry tab section that includes premitigated risks data. On the lower section of the risk data entry tab users may enter the postmitigated risks data. The table header presents short titles of the data entered in respective columns. The dark, shaded cells represent the minimum data needed to be entered in order for the model to run. These shaded cells require minimal knowledge of risk analysis and may be used for projects where risks are independent and noncorrelated.

TABLE 6.4   Risk Data Entry Sheet (tab or table)

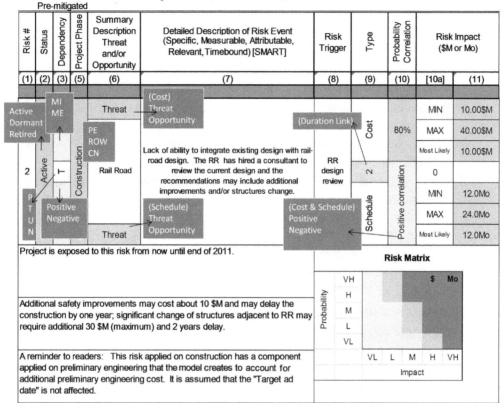

For more advanced risk analysis (which may consider risk conditionality), users need to understand and use correctly the lightly shaded cells. The next few paragraphs exemplify each cell's functionality. The gray boxes indicate the content of drop-down menus. This feature allows users easy data entry in the correct cell. Each cell has a validation pop-up window that presents information about that cell's functionality.

Detail Description

1. **Risk #** indicates the position of risk among risks assessed. It does not relate with its magnitude or criticality. The number coincides with the number of the risk data entry tab.

2. **Status** indicates the risk's state: (1) active—when the risk is being actively managed or monitored, (2) dormant—when the risk is real but the project management chose not to spend time on it, and (3) retired—when the risk is terminated by response action or because it did not materialize. When the status of a risk is "retired" the risk does not contribute to the calculation. It is recommended that after a while the risk be taken out

and archived. The first two states (active and dormant) participate in the computation with their full impact.

3. **Conditionality** has three data entry fields:

   3.1. The upper section captures risk dependency on the "above" risk event. The drop-down menu allows selection of: (a) "MI" for mutually inclusive risks, (b) "ME" for mutually exclusive risks, and (c) blank for independent risks.

   3.2. The central cell indicates the type of distribution assigned to risk. The current version of RBES takes only a Pert distribution, which is the default distribution. Newer versions will include distributions such as: triangular, normal, and uniform in addition to the Pert distribution.

   3.3. The lower section indicates the correlation type with the previous risk. The drop-down menu allows selection of positive, negative, or blank, which indicates that the risk's distribution is positive correlated, negative correlative, or noncorrelated with the previous risk.

4. **Project phase** indicates which project phase is affected. The drop-down menu allows three options: (1) preliminary engineering (PE), (2) cost of land use (ROW), and (3) construction (CN).

5. **Risk name (threat or opportunity)** has three sections:

   5.1. The upper section indicates if the risk's cost component is a threat or opportunity. It is important to select the right type of risk because the model treats opportunities in a manner of savings. The quantitative impact always is entered with positive values and the model will assign a negative sign to the opportunity so the opportunity's impact will reduce the overall cost.

   5.2. The central cell shows the risk's name. Risk names should be short but descriptive.

   5.3. The lower section indicates if the duration component of the risk is a threat or opportunity. It is important to select the right type of risk because the model treats opportunities in a manner of reducing an activity's duration. The quantitative impact always is entered with positive values and the model will assign a negative sign to the opportunity so the opportunity's impact will reduce the overall activity duration.

   The reader may have noticed that RBES allows different combinations for the cost and schedule. Table 6.5 shows the four types of combinations possible for cost and schedule risks' components.

6. **Detail description of risk events** is the place to record information about risks. The information needs to be recorded in a clear and concise manner. The SMART principle

TABLE 6.5   Possible Combination of Threat or Opportunity for Cost and Schedule

	1	2	3	4
Upper cell	Threat	Threat	Opportunity	Opportunity
Lower cell	Threat	Opportunity	Threat	Opportunity

(see Table 3.6) is the best way to develop a good risk description. It is important to present all assumptions made and the description needs to make sense to a layperson.

7. **Risk trigger** presents the indicators (symptoms) that show that the risk is about to materialize. Risk trigger is an excellent monitoring element that the project manager and risk manager use in their efforts of controlling its effects.

8. **Type** has three sections:

    8.1. The upper section indicates that the risk impact values presented in column (11) relates to the cost impact. Every risk is presented through its two components: cost, on upper section, and schedule, on lower section of risk data entry table.

    8.2. The central cell reflects an advanced feature of RBES that allows the modeler to capture situations of two schedule risks that are applied on the same activity and are in series. (See Chapter 4—Schedule Risks.) The odd-numbered schedule risks may be assigned to be in a series with the next schedule risk. Figure 6.5 presents how and what the modeler has to enter in order to capture two series risks.

    The first risk in the series is called "master duration risk" and the modeler enters "1" in its specified cell (the central cell that the arrow points out) and in the next risk the equivalent cell will appear as 1, which indicates that the duration of these two

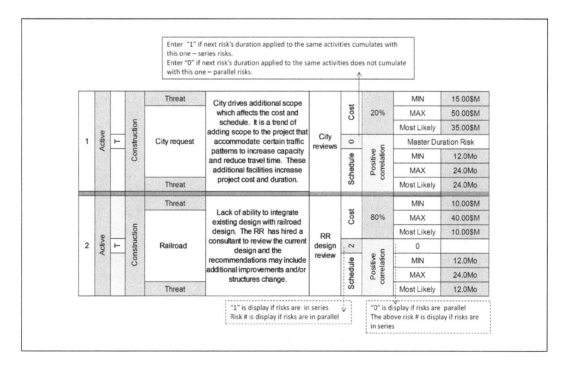

Figure 6.5 How to Enter Schedule Risks: Series or Parallel

risks are in series (they cumulate). The cell just above "MIN" for the schedule will present the risk number of the above risk.

When risks are in parallel, the master duration risk may have the default value (blank), or select from the drop-down menu "0" and the next risk will display in the equivalent cell its own risk number. The cell just above "MIN" for the schedule risk will display "0."

8.3. The lower section indicates that the risk impact values presented on column (11) relates to the schedule impact.

9. **Probability/correlation** has two sections:

9.1. The upper section is used to enter the risk's probability of occurrence. The user may enter any value from 0 to 100 percent, although remember not to select numbers such as 52 or 81, because this implies false precision as discussed in Chapter 3.

9.2. The lower section indicates correlation that might exist between risk's cost and schedule distributions. A drop-down menu allows selection of: (1) positive correlation, (2) negative correlation, and (3) noncorrelated (blank option.)

10. & 11. **Risk impact** has three major sections:

11.1. The upper section captures the cost risk impact (effect) values when risk occurs. All values are positive and in the case of threats, they represent the additional values on top of the estimated base cost. In the case of opportunity, they represent savings (reduction in cost) that are taken out of the estimated base cost.

11.2. The middle section is connected to the schedule risks presented in Section 8.2.

11.3. The lower section captures the schedule risk impact (effect) values when risk occurs. All values are positive and in the case of a threat, they represent the additional values on top of the estimated base duration. In the case of an opportunity, they represent time savings (reduction on duration) that are taken out of the estimated activity's base duration.

The lower side of the table allows users to record additional information about the risk. We recommend that any assumption made be recorded and any suggestion regarding a response to the risk be clearly presented. The user may customize each of the three cells based on the project's needs.

The risk matrix display is automatically refreshed by activating a macro bottom.

The risk data entry tab has two similar tables as described previously: one for the premitigated (original) scenario and the second one for a postmitigated scenario (after the project management defines a risk management plan and the new risk mesh will has a different configuration).

The authors state again that the quality of RBE depends on the quality of data used. It is critical to have identified and quantified the significant risks (threats and opportunities). At the same time, a clear understanding of a project's risk mesh is crucial. If risks' conditionality is not captured properly, the risk analysis fails.

RISK CONDITIONALITY AND RBES

Risk conditionality is a segment of risk assessment that brings specificity to the project risk mesh. The majority of project risks are independent and noncorrelated, and because of that the default of any risk analysis considers risks independent and noncorrelated. This default approach (independent risks and no correlation among them) produces an undesirable trend in RBE. The authors have noticed too many times improperly defined project risk mesh because of ignoring (skipping) the identifying and recording risk conditionality when it is warranted.

Understanding Risk-Based Relationships

Risks and base cost and/or schedule work together through MCM simulation. While MCM is a simple statistical analysis method the interpretation of its results may puzzle some. Assuming that the results are correct, the risk analyst must understand them and be able to answer any questions that may come.

The challenge intensifies when quality assurance/quality control (QA/QC) is performed. The QA/QC requires good understanding of risks' role in analysis results. The QA/QC is an essential task of RBE that may be performed in different ways, depending on how your organization is set up. Regardless of QA/QC technicality, the RBE results must pass the "common sense" test. If this test is difficult to pass, then the QA/QC has no value.

The common sense test is nothing more than having unambiguous answers to reasonable questions that may arise. Frequently asked questions include the following:

- Why are the maximum cost or schedule values so high?
- Why does the distribution have such long end tails?
- Why the double hump distribution?
- What is in humps?
- What is in dips?
- Why triple hump distribution?

The ability of dealing with the common sense test improves once the risk analyst understands the basics of results' interpretations. The next few paragraphs introduce the basics of how risk and base cost and schedule are reflected in risk analysis results. How a risk (cost and schedule) and base (cost and schedule) interact are presented and illustrated through data entry and simulation results. The following examples are designed to build skills and ability for quality common sense tests.

The simplest situation is when the base cost and schedule have their own variability ($\pm$ 10%) and risk applied to them is a perpetual one (100 percent probability of occurrence). Figure 6.6 presents screenshots of RBES that illustrate the quantitative risk values (threat, in this case) for cost and schedule when the risk is perpetual. The pie diagram is quite simple because there is just one possibility (base and risk always occur simultaneously).

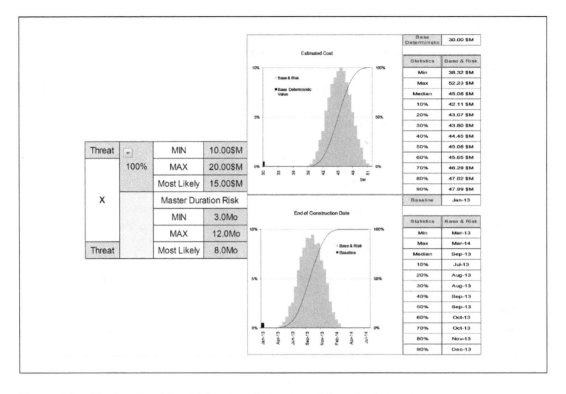

Figure 6.6   The Results of Combining Base Estimate and Perpetual Threat

All examples presented assume that the base cost is $30 million with a variability of ±10 percent. The schedule has a baseline value of January 2013 as end of activity, where the activity duration is 24 months. A variability of ±10 percent is applied to the activity's duration. Risks are defined by their quantitative values using the risk data entry format. As a reminder, the upper section of the risk entry table represents risk's cost component and the lower section represents the risk schedule component. For communication reasons, risks are tagged with names such as "X" and "Y."

The upper diagram represents results of cost analysis and the lower represents schedule analysis in forms of a histogram and a cumulative distribution function. The base's most likely value is plotted to emphasize the risk's threat or opportunity characteristics. Since the risk has 100 percent probability of occurrence the results show only distribution of base plus risk together.

On the right side of Figure 6.6 (and the following figures) the tabular percentiles are presented to allow readers to become familiar with the most common ways of presenting RBE results. The activity analyzed is called construction and its base data are presented by their most likely value of cost estimate and most likely value of end of construction date.

Figure 6.7 presents a similar situation as that presented in Figure 6.6, with the only difference being the fact that the risk is now an opportunity.

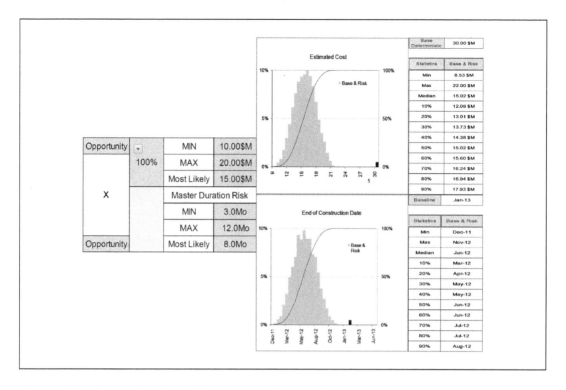

Figure 6.7   The Results of Combining Base Estimate and Perpetual Opportunity

Figure 6.7 shows in a clear manner how RBES treats opportunities. The entry data dictate how much the opportunity will reduce the cost or will shorten the activity's duration. The numbers entered are positive and RBES recognizes the fact that the risk is an opportunity and subtracts the opportunity's impact from the base cost or base duration. Figure 6.7 shows that the estimated cost is lower than the base cost and that the estimated end of construction is sooner than the baseline schedule when risk occurs.

The next two figures present the situations of having a *no clue risk event* when its probability of occurrence is 50 percent. Figure 6.8 is having both components as opportunities and Figure 6.9 is having cost risk as a threat and schedule risk as an opportunity.

The risk's 50 percent probability of occurrence imposes that the horizontal segment of CDF occurs at 50 percent on the secondary axis and the area covered by each hump of the histogram is equal. Since the range of the base variability is smaller than the range of combined base and risk impact, the base cost hump is taller than the base and risk hump.

Figure 6.9 shows the situation when the base cost and baseline duration have deterministic values (just one number). In this way the base is represented by one single bar that rises up to 50 percent on the primary axis. The figure allows viewers to see clearly that 50 percent of the time the model picks only base value (risk doesn't occur) and the remaining 50 percent adds to the base values the risk impact values.

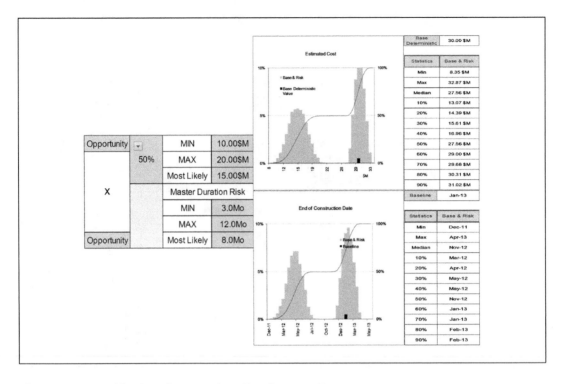

Figure 6.8   Combination of Base and No Clue Opportunity

Relationship between Two Risks

Different situations will be presented and discussed regarding the relationship between events and the correlation of their impact.

Each situation presents the following:

1. Quantitative impact of each risk associated with their conditionality
2. Base cost uncertainty
3. Results in graphic form (histogram and cumulative distribution function)
4. Results in percentile table
5. Probability analysis (probability tree) with risk's probability of occurrence of each possible combination
6. Probability of occurrence of each possible combination (pie chart)

The situations are grouped from simple to complex with "no clue event probability of occurrence" of each risk. There are many possible situations and we cannot cover all of them; however, you are encouraged to examine each situation presented and then practice with your own data.

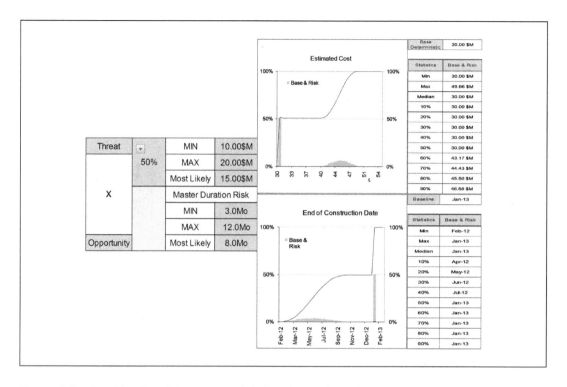

Figure 6.9   Combination of Base Deterministic Value and No Clue Events

Independent Risks Only

Figure 6.10 represents the situation of two risks that are independent of each other as events and distribution (noncorrelated). The probability tree and pie chart show that the probability of occurrence of each possible combination of risk and base is 25 percent. This number will change when risks' probability of occurrence change. Base cost zero is represented by a single bar that reaches 25 percent likelihood on the primary axis. The area covered by each hump of the histogram is the same and equals 25 percent likelihood.

Figure 6.11 represents the same scenario as the previous one but now the base has a value of $30 M and variability of ±10 percent. Each hump covers the same area. The high of each hump is in inverse relationship with the hump's base so the hump representing the base alone is taller because the base distribution given by its variability has the narrowest range.

Mutually Inclusive Risks

The code "MI" is entered in the box, as shown in Figure 6.12. Y is 100 percent mutually inclusive with X and it means that risk Y occurs *only and always* when risk X occurs. In other words, Y is totally mutually inclusive with X. The outcomes of this situation consist of a histogram with two humps. The horizontal segment of CDF is at the 50 percent level of the secondary axis. If the probability of occurrence of X is changed at 30 percent, the horizontal segment of CDF will move

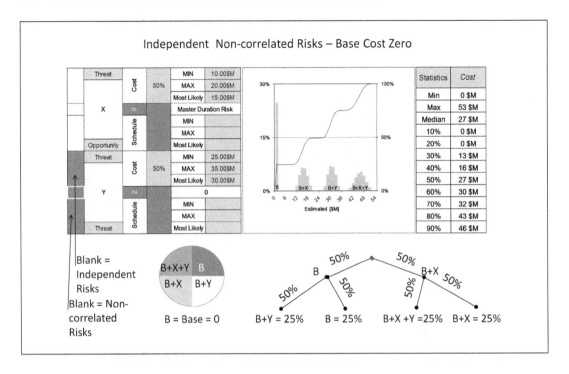

Figure 6.10   Base Zero and Two Independent Noncorrelated Risks

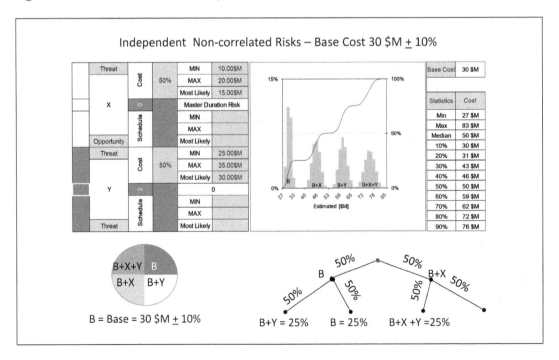

Figure 6.11   Base Variable and Two Independent Noncorrelated Risks

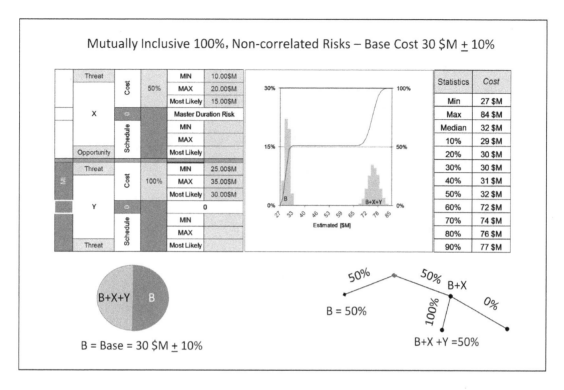

Figure 6.12   Base Variable and Two Mutually Inclusive 100 Percent Threats

at the 30 percent level of the secondary axis. Since both risks are threats the second hump is definitely to the right of the hump representing base only.

The situation is different if we change the mutually inclusive relationship from total mutually inclusive to partial mutually inclusive. Figure 6.13 shows the situation when Y may occur only half of the time when X occurs. In this case the pie chart has three sections representing all possible combinations.

Mutually Exclusive Risks

The next four figures present mutually exclusive risks in different situations. Figure 6.14 shows the situation when the second risk occurs *only and always* when the *first risk does not occur*. It represents the total mutually exclusive risks when the occurrence of one risk excludes the other risk. There are only two possible combinations when base can be associated with X or with Y.

Figure 6.15 shows the situation when Y has a 50 percent chance to occur *only when X does not occur*. It is called a partial mutually exclusive relationship. In this case there are three possible combinations presented by the pie chart and probability tree.

Mutually exclusive risks in No Clue Events are presented in Figure 6.16. This type of relationship is created when risk elicitation defines two mutually exclusive risks that have the same chance of occurrence as base alone. Each of the combinations has approximately a

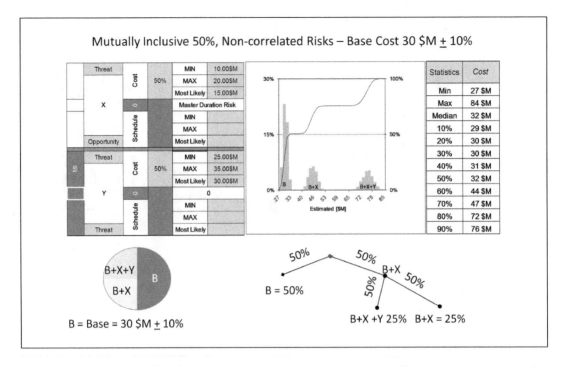

Figure 6.13 Base Variable and Two Mutually Inclusive 50 Percent Threats

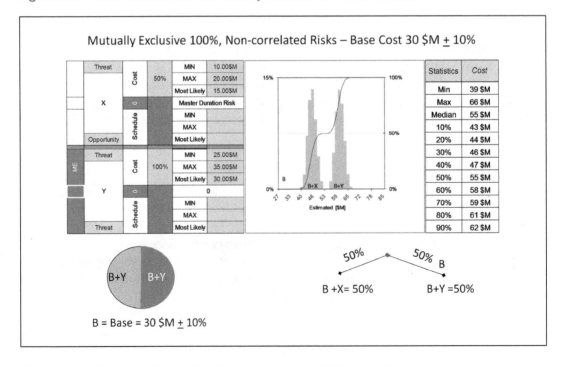

Figure 6.14 Base Variable and Two Mutually Exclusive 100 Percent Threats

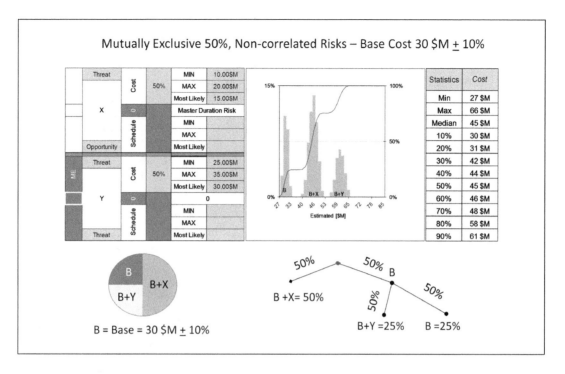

Figure 6.15 Base Variable and Two Mutually Exclusive 50 Percent Threats

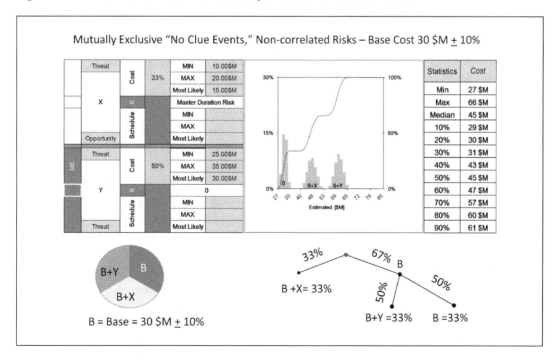

Figure 6.16 Base Variable and Two Mutually Exclusive No Clue Threats

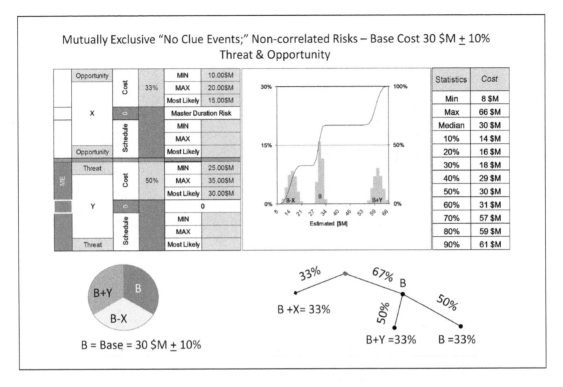

Figure 6.17   Base Variable and Two Mutually Exclusive No Clue Threat and Opportunity

33.3 percent chance of occurrence. This situation is captured by assigning to one risk a 33.33 percent probability of occurrence and imposing a mutually exclusive relationship to the second risk and 50 percent of occurrence of remaining the 67.67 percent when the first risk does not occur. The horizontal segments of CDF are at even intervals along the vertical axis.

Figure 6.17 shows a similar situation with one risk being an opportunity and the other a threat. The base-only hump is between the other two humps, which represent base and opportunity and base and threat, respectively.

For comparison's sake, Figure 6.18 presents the same base and risks as Figure 6.17 but with the difference that Figure 6.18 shows two independent risks and Figure 6.17 shows two mutually exclusive risks. In both situations risks have no clue events as their probability of occurrence. The figures show clear distinction between results. As expected, a new hump is created by the combination of base and both risks together.

Correlation Analysis

The next six figures present the effect of risks' correlation over analysis results. Since correlation makes sense only when risks occur the examples presented consider total mutually inclusive risks. The first three figures show the situations when base cost has significant value ($30 M ±10 percent) and the second group of three figures shows the situations when base cost is zero.

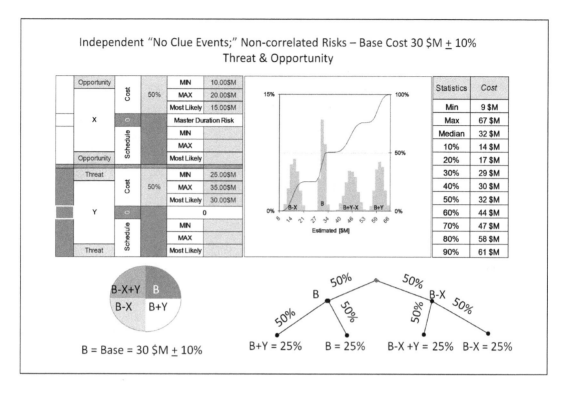

Figure 6.18   Base Variable and Two Independent No Clue Threat and Opportunity

Figure 6.19 presents the noncorrelated risks scenario, Figure 6.20 presents the positive correlated risks, and Figure 6.21 presents negative correlated risks. The mutually inclusive code and 100 percent probability of occurrence of the second risk indicate that the second risk (Y) occurs only and always when the first risk (X) occurs.

A quick examination shows that the largest hump range occurs when risks are positive correlated; noncorrelation has a narrower range, and finally, the narrowest hump range is provided by the case of risks in negative correlation. The reader may notice that when risks are negative correlated (Figure 6.21) the base-only and base-plus risks humps have the identical shapes.

The percentile tables present in numeric form the correlation effect over results. Since both risks are threats the result minimum values is unchanged. Furthermore, since the first risk (X) has 50 percent probability of occurrence, the percentiles values up to 50 percent are identical. The effect of correlation starts to kick in after 50 percent. At 50 percent positive correlated risks has the lowest value and negative correlated risks has the highest value. This trend is reversed at 75 percent percentile level. At this level, all three cases (noncorrelated, positive correlated, and negative correlated) have the same value ($90 M). Above the 75 percent level the positive correlated risks has the highest value and negative correlated risks has the lowest value.

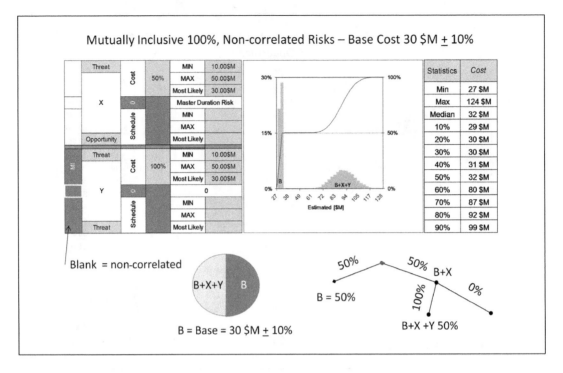

Figure 6.19   Base Variable and Two Mutually Inclusive 100 Percent, Noncorrelated Threats

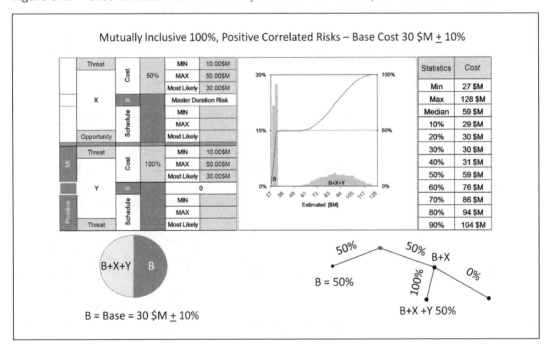

Figure 6.20   Base Variable and Two Mutually Inclusive 100 Percent, Positive Correlated Threats

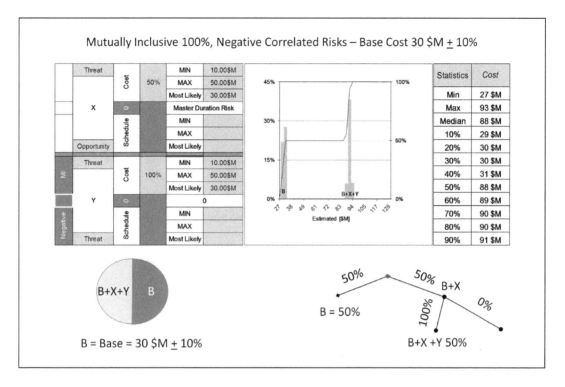

Figure 6.21   Base Variable and Two Mutually Inclusive 100 Percent, Negative Correlated Threats

The next three figures present similar situation as the previous three figures but they display different results because the base cost is zero. Figures 6.22 to 6.24 represent the effect of correlation upon two perpetual risks. When risks are perpetual, it does not matter if they are considered mutually inclusive or independent since both always occur.

The effect of correlation on risk analysis results is clearly illustrated by the histograms, CDF, and percentiles tables presented in the previous figures. Regardless of the type of correlation all results have the same mean, and in our situations the mean coincides with the median (symmetrical distribution). The noncorrelated risks (Figure 6.22) have the histogram higher than the positive correlated risks (Figure 6.23) and the range of distribution narrower than positive correlated risks. Figures 6.22 and 6.23 show that correlation has a significant effect on the analysis outcome.

The negative correlation offers only one single number for results. It is the sum of the mean values. Figure 6.24 offers a powerful example of how correlation works. Readers may be surprised to learn that combining two distributions (it does not matter how large a range they may have) can lead to one single number.

The positive correlation between risks expands the tails of combined distribution, while negative correlation between risks shrinks the range of combined distribution. Warning: Correlation

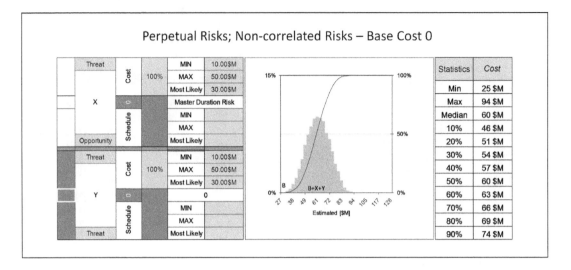

Figure 6.22   Two Perpetual Threats; Noncorrelated

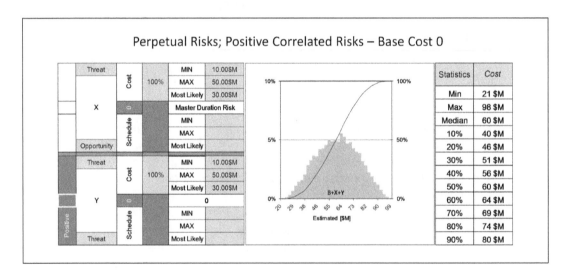

Figure 6.23   Two Perpetual Threats; Positive Correlated

among variables must represent actual conditions. Any abuse of unjustified use of correlation disturbs the analysis and ultimately diminishes credibility in RBE.[5]

Schedule Risks

The estimated cost is calculated for each of the iterations by algebraic addition of base cost values and risks cost values that the model extracts from each distribution. The schedule is a

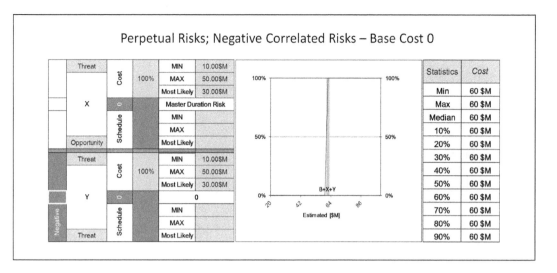

Figure 6.24   Two Perpetual Threats; Negative Correlated

little different since the model must consider critical path (see the section "Schedule Risks" in Chapter 4). For example, if two schedule risks occur on the same activity it may produce two distinct situations: (1) the activity duration is affected by the longest schedule risk (the schedule risks are in parallel), or (2) the activity duration is affected by the cumulative risks values (the schedule risks are in series).

The next two figures present two risks with identical schedule components that are in parallel (Figure 6.25) and series (Figure 6.26.) Figure 6.25 presents details of how the data must be entered and what the user should expect at check boxes. The schedule risks have the same probability of occurrence (our favorite no clue events) and the same impact distribution.

Figure 6.25 shows two humps: (1) base-only impact and (2) base-plus risk X, or Y, or X and Y. Since risks impact have the same value the risks effect is represented by one single hump. When both risks occur, the model picks up the longest one. Figure 6.26 shows that changing from parallel risks to series risks introduces the third hump because when both risks occur the model will add the X and Y values to the base value determining the activity duration for that iteration.

The next two figures present two risks that have significant different distributions. Risk Y is approximately three times lower than X. Both figures show three humps but their significance is different. Figure 6.27 represents the situation when schedule risks are in parallel. Since the schedule risk X is larger than schedule risk Y, the model will select X values when both risks occur because X is greater than Y. This explains why risk X dominates the distribution.

Figure 6.28 shows the same conditions as Figure 6.27 with a changed relationship from parallel to series. The results are changed at the higher end of the distribution since the hump at the right has a wider range. That hump includes values when risk X occurs alone and when risks X and Y occur simultaneously. When both risks occur, the model adds their values and this supplemental risk increases the hump's range.

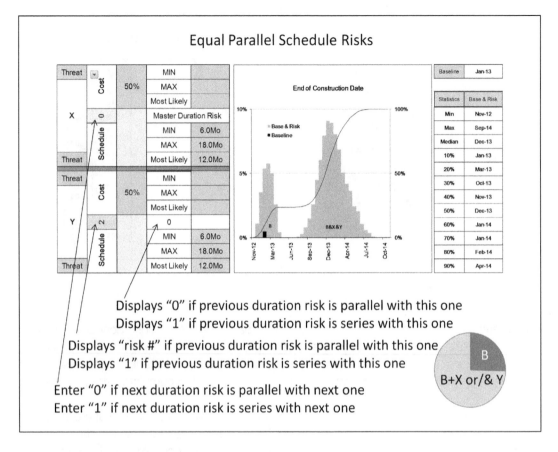

Figure 6.25 Two Identical Parallel Schedule Threats

Summary

Risk-based estimate self-modeling spreadsheet facilitates cost and schedule risk analysis by allowing users with minimal risk analysis experience to enter data in the model and run simulations. The RBES offers convenient and comprehensive templates waiting to receive data regarding the base cost and schedule and risks to them. Risks' conditionality may be captured in their full meaning and users must be certain that the conditionality entered (even default ones) is accurate.

MODEL OUTPUTS

The Estimated Cost of the Project Main Phases and Total

The micro-project defined in Chapter 3 presents how RBE may show its cost results (see Figure 3.36, Table 3.10). The results presented in Chapter 3 were computed using an off-the-shelf

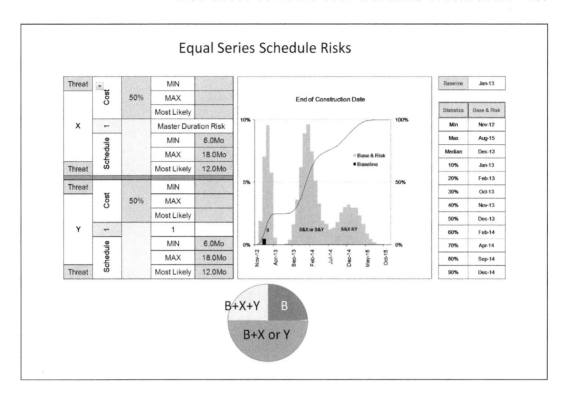

Figure 6.26   Two Identical Series Schedule Threats

software program and considered all variables, as uniform distribution and risks were unaltered by any markups.

The same micro-project is going to be analyzed using project-like conditions such as: (1) variables are represented through symmetrical Pert distribution and (2) the construction risk is affected by risks' markups. The micro-project is cost-only analysis so only a part of RBES futures is employed. The data are entered into the base tab and risks tabs and then moved into the RMP tab. Figure 6.29 shows a screenshot of RMP base cost–related data.

Since the micro-project is a cost-only estimate the information related to project schedule may take any values and users will consider only current year cost results ignoring results of year of expenditure or schedule. In order to emphasize the lack of schedule information (cost-only risk analysis) the RBES considers the same date for "Estimate Date" and "Target Ad Date" and 0.0001 month for construction duration. The base cost for PE, ROW, and CN is the median of values provided in Chapter 3 and their variability is calculated based on the distribution range presented in Chapter 3 as well.

The micro-project has three risks, with each of them affecting just one activity. Figure 6.30 shows a screenshot of the RMP risks area limited to quantitative data.

You may notice that the risks are independent and noncorrelated as assumed in Chapter 3. It is quite simple data input. Running 10,000 iterations produces results that are displayed in

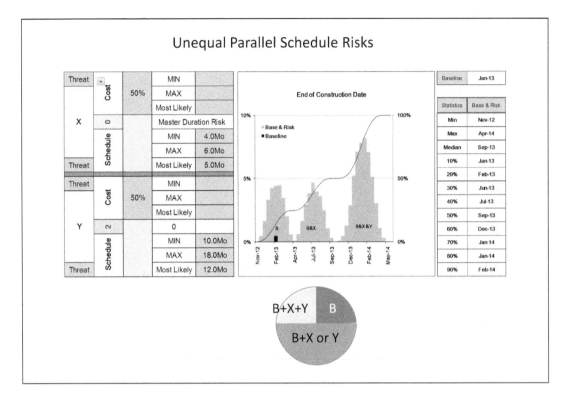

Figure 6.27   Two Distinct Parallel Schedule Threats

Figure 6.31. The second hump on construction cost estimate distribution is prominent. Overall, the comments made in Chapter 3 are valid here as well and the cost distributions of PE, ROW, CN, and Total have better resolution now because of employing Pert distribution as defining the variable's shape.

Candidates for Risk Response (Tornado Diagram)

Risk-based estimate self-modeling provides the candidates for risk management as defined in Chapter 3. Since the micro-project has only three risks the candidates for risk management are quite simple and intuitive. Figure 6.32 confirms what the reader may have anticipated.

RISK MANAGEMENT

One of the most important benefits of RBE consists of being an integral part of the project risk management process, as it was discussed in previous chapters. RBES offers plenty of opportunities of capturing the assumptions and decisions related to risk management. Risk data entry tabs allow users to document a risk event and capture information regarding risk response planning.

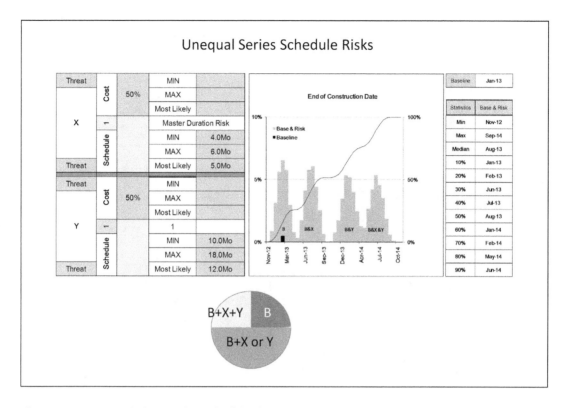

Figure 6.28   Two Distinct Series Schedule Threats

Project Title		Micro-project	Value	Variability	Risk Markups	
Estimate Date	01/12/10	Target Ad Date	01/12/10	0%	Mob	10.0%
Project PIN #		Estimated CN Duration	0.0Mo	0%	Tax	9.0%
Last Review Date		Estimated PE Cost	3.00 $M	17%	CE	12.0%
Project Manager		Estimated ROW Cost	12.00 $M	8%	PE	9.0%
		Estimated CN Cost	33.50 $M	4%	C.O.C	4.0%

Figure 6.29   Micro-Project Base Data and Risks' Markups Sheet

1	Active		Pre-construction	Threat	Cost	50%	MIN	0.50$M
							MAX	0.80$M
							Most Likely	0.65$M
				PE	0		Master Duration Risk	
					Schedule		MIN	
							MAX	
							Most Likely	
2	Active		ROW	Threat	Cost	30%	MIN	1.00$M
							MAX	5.00$M
							Most Likely	3.00$M
				ROW	2		0	
					Schedule		MIN	
							MAX	
							Most Likely	
3	Active		Construction	Threat	Cost	67%	MIN	2.00$M
							MAX	8.00$M
							Most Likely	5.00$M
				CN	0		Master Duration Risk	
					Schedule		MIN	
							MAX	
							Most Likely	

Figure 6.30 Micro-Project Risk Data

Risk's Qualitative Matrix Display

The risk management plan (RMP) is designed to bring all base estimate and risks information on the same piece of paper so the reader may have a better view of project risks. The RMP tabs include for each risk the qualitative display of the risk expected impact value (distribution's mean value) and its probability of occurrence as shown in Figure 6.33.

Risk's matrix, shown in Figure 6.33, indicates that the risk's probability of occurrence is moderate and the risk's cost component has a very high impact value while its schedule component has only a moderate impact value. This risk deserves full attention because of its

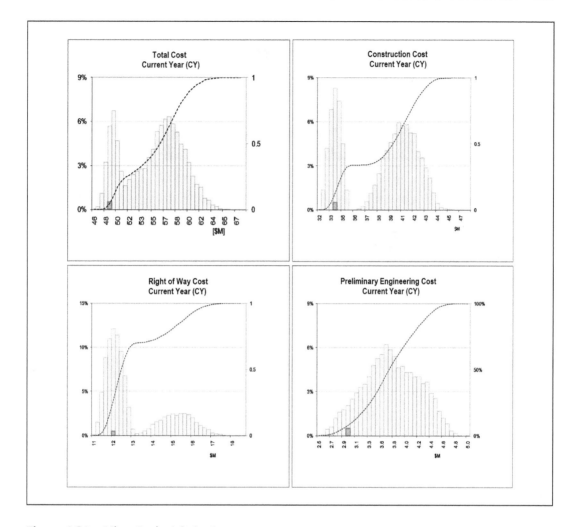

Figure 6.31   Micro-Project Outputs

cost component. A response plan may affect probability of occurrence, cost impact values, and schedule impact values.

Figure 6.34 shows the criteria used to transfer risk's quantitative values into risk qualitative display data. The left column of Figure 6.34 indicates the probability of occurrence criterion of defining the qualitative scale. The criterion may be easily adapted to any specific project's owner requirements. We have chosen a proportional scale, but for some projects it may make better sense to use a different scale.

The qualitative risk's impact terms (for the cost and schedule) are defined in relation to construction base cost and total project base duration as shown in the middle column (cost)

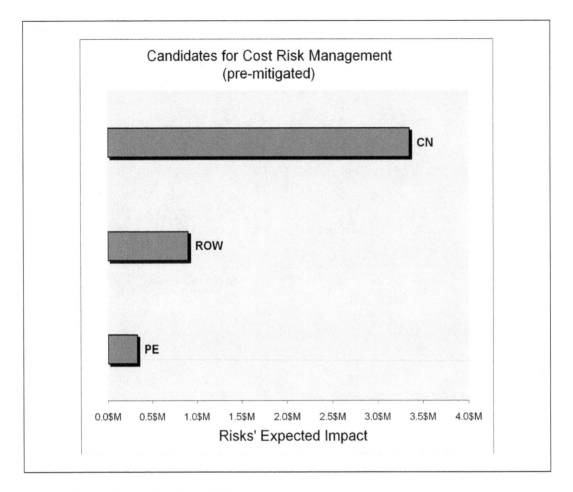

Figure 6.32   Micro-Project Tornado Diagram

Cost	50%	MIN	15.00$M	18.79$M		Very High	VH		
		MAX	50.00$M		Moderate		H		
		Most Likely	35.00$M				M	Mo	$
Schedule	0	Positive correlation	Master Duration Risk			Moderate	L		
							VL		
		MIN	2.0Mo	3.21$M					
		MAX	9.0Mo				VL L M H VH		
		Most Likely	6.0Mo				Impact		

Figure 6.33   Quantitative Risk Data and Qualitative Display of Risk's Mean Impact Value

Probability			Cost risk impact			Duration risk impact	
Very High	80%		Very High	>>> 0.1*L6		Very High	>>> 0.3*(L$3+(L$2-F$2)/30.5)
High	60%		High	>>> 0.05*L6		High	>>> 0.2*(L$3+(L$2-F$2)/30.5)
Moderate	40%		Moderate	>>> 0.02*L6		Moderate	>>> 0.1*(L$3+(L$2-F$2)/30.5)
Low	20%		Low	>>> 0.008*L6		Low	>>> 0.05*(L$3+(L$2-F$2)/30.5)
Very Low	0%		Very Low	>>> 0.0005*L6		Very Low	>>> 0.01*(L$3+(L$2-F$2)/30.5)

L6 >>> Construction Cost

L$3 >>> Estimated CN duration

L$2 >>> Target Ad date

F$2 >>> Estimate date

Figure 6.34   Transition Factors from Quantitative to Qualitative Risk Data

and the right column (schedule) of Figure 6.34. The factors chosen depend on the values of risks identified. Each project may adopt its own criterion in order to emphasize some hierarchical display. In other words, the display should avoid having all risks showing very high impact values or very low impact values. A good display should cover in a balanced way the entire spectrum (from very high to very low) of qualitative impact values.

Risk Response Plan

The risk management plan (RMP) tabs include sections on the right side of the spreadsheet that facilitate risk response planning, monitoring, and control. Figure 6.35 shows how this section is organized. The first column (16) allows users to choose the risk response strategy applied in order to ameliorate or terminate the threat or to enhance or increase the chance of opportunity. (An in-depth discussion of the advantage or disadvantage of any strategy adopted was presented in Chapter 5.)

Column (17) of Figure 6.35 brings specifics about the actions to be taken. It is important to include all pertinent information about the actions taken since it documents the effort of risk response.

Risk Monitoring and Control

Column (18) in Figure 6.35 indicates the person or entity with direct responsibility of managing that risk. Column (19) shows the dates when risk was reviewed and column (20) includes statements about what happened and how decisions were made.

	Risk Response Plan		Monitoring and Control			Critical Issue
Strategy	ACTION TO BE TAKEN Response Actions including advantages and disadvantages, include date	Risk Owner	Risk Review Dates	Date, Status and Review Comments (Do not delete prior comments, they provide a history)		Is Risk on Critical Path?
(16)	(17)	(18)	(19)	(20)		(21)
Mitigation	Finalize design to identify all wetlands that are impacted. Early coordination with the outside agencies to determine mitigation ratio.	Design Leader/Enviro. mgr	2007-Jan-2 2006-Dec-2	As of Nov. 15, 2005 there are only two potential areas where there could be additional wetland impacts. As of Dec. 2, 2005 agency has initially determined that mitigation ratio would be 4:1.		YES

Figure 6.35   Risk Response and Monitoring and Control Section

Project Title		Eastside Program -- Section A		Value	Variability	Risk Markups		Inflation tables built-in.	
Estimate Date	04/20/10	Target AD date		07/01/11	10%	Mob	10.0%	A/B/A Duration	1Mo
Project PIN #		Estimated CN Duration		101.0Mo	10%	Tax	8.7%	Annual inflation rates	YOE
Last Review Date		Estimated PE Cost		20.54 $M	10%	CE	10.0%	PE	1%	20.7$M
Project Manager	JD	Estimated ROW Cost		74.81 $M	10%	PE	10.0%	ROW	2%	75.7$M
		Estimated CN Cost		205.40 $M	10%	C.O.C	4.0%	CN	3%	241.1$M

Inflation Points		Base Cost Market Conditions			
Define inflation point of the activity cost. For example 50% means that the inflation point for that activity is the mid-point activity. 50% is the default value. If it is decided that the inflation point is at three quarters of respective activity then you must enter 0.75		Better than planned	10%	10%	**Update the pre-mitigated base cost and risks data**
		Worse than planned	20%	30%	
		Probability	Impact		
Preconstruction activities (ROW and PE)	0.8				
Construction					

Figure 6.36   Section A—Base Estimate Data Entry

Usually RBE provides guidance on possible risk response plans, but the main responsibility on risk management belongs to the project manager. The boxes provided on RBES are supposed to be completed by the project manager and other project team members. The risk management plan introduces changes to the project base estimate and schedule and changes regarding risks' values. When these changes are significant, new simulations should be run to obtain the postmitigated estimate results. These results are the values for project budgeting.

CRITICAL ISSUES

Column (21) of Figure 6.35 indicates if schedule risk is on critical path or not. It allows two options: (1) "YES" when risk is on critical path and its value affects schedule results, and (2) "NO" when schedule risk does not belong to critical path and in this condition its schedule value is ignored. At the same time, the data in column (21) may capture other issues not yet specified.

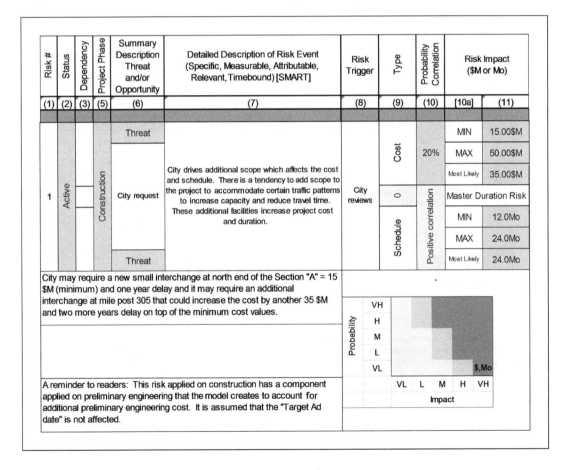

Figure 6.37   Section A—Risk 1

TUTORIALS/CASE STUDIES

Case Study No. 1: Eastside Corridor — Section A

Eastside Corridor was introduced in Chapter 3 as an example of the versatility of RBE. Chapter 3 indicated that the cost and schedule risk analysis needs to be performed in two stages: (1) section "A" and (2) corridor level, to accommodate the available project data and the project owners' request. Chapter 3 explains the rationale and methodology applied and the next paragraphs present the main steps and screenshots of data entry and results regarding section A cost and schedule risk analysis.

Based on available data the base cost and schedule team has reviewed and validated the project team estimate. The results of the validated base estimate are entered into the base tab of RBES, as shown in Figure 6.36. A 10 percent variability of the base has been adopted for

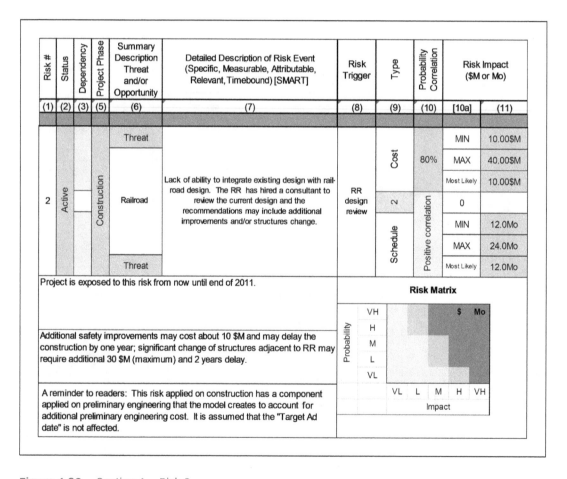

Figure 6.38   Section A—Risk 2

all base values. The risk markups coefficients were taken from the validated base cost estimate and entered in base tab.

The workshop has determined the following inflation rates: (1) 1 percent for preliminary engineering, (2) 2 percent for land acquisition, and (3) 3 percent for construction cost. The inflation is considering that money for land acquisition is spent closer to the end of land acquisition activity (0.8) while for construction, a uniform spending is assumed (blank or 0.5).

The market conditions entered recognize the possibility of having bids lower than validated base cost—by 10 percent of base most likely value or higher than validated, by 30 percent of base most likely value. The chance of having worse than planned market conditions is two times larger than the chance of having better than planned.

By clicking the macro button "Update the premitigated base cost and risks data," all data entered in the base tab and risk tabs are transferred to the RMP and RMPSuppl tabs.

Figures 6.37 through 6.46 present the information about risks identified and quantified through advanced risk elicitation interviews and workshops. Three pairs of risks (3 and 4;

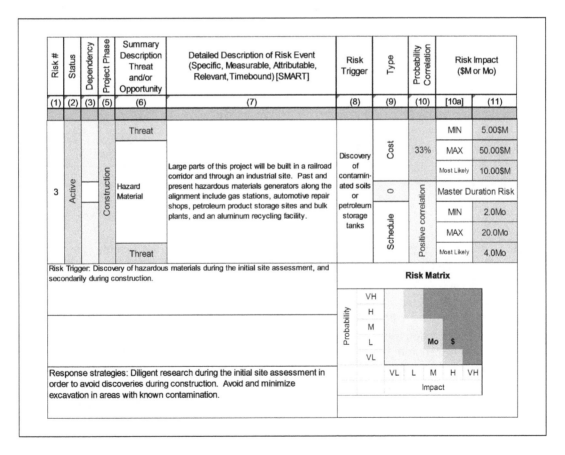

Figure 6.39   Section A—Risk 3

Risk #	Status	Dependency	Project Phase	Summary Description Threat and/or Opportunity	Detailed Description of Risk Event (Specific, Measurable, Attributable, Relevant, Timebound) [SMART]	Risk Trigger	Type	Probability Correlation	Risk Impact ($M or Mo)	
(1)	(2)	(3)	(5)	(6)	(7)	(8)	(9)	(10)	[10a]	(11)
				Opportunity			Cost	50%	MIN	1.00$M
									MAX	4.00$M
									Most Likely	3.00$M
4	Active	ME	Construction	Hazard Material 2	The base cost for hazardous materials—cleanup and discovery may be more than sufficient. It may be an opportunity of avoiding some of its cost.		4		0	
							Schedule		MIN	
									MAX	
				Opportunity					Most Likely	

Note: Preliminary engineering cost created by the default of the model needs to be disable.

Risk Matrix

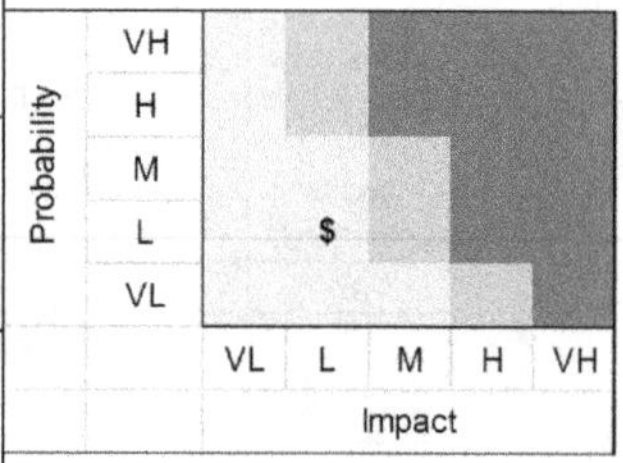

Figure 6.40   Section A—Risk 4

5 and 6; 9 and 10) brings specificity to the project's risk mesh. The dependencies of these pairs of risks are partial mutually exclusive relationships.

Figure 6.47 shows how the hazardous materials may affect the project cost and schedule. The base cost includes $8 M to cover the cost of cleaning up the hazardous materials that are on the construction site. Large segments of this project will be built in a railroad corridor and through an industrial site. Past and present hazardous material generators along the alignment include gas stations, automotive repair shops, petroleum product storage sites and bulk plants, and an aluminum recycling facility.

How much hazardous material is on the site is anyone's guess. At the time of the workshop (estimate) it was assumed that $8 M may cover the cost of cleanups but it was recognized that cleanups may produce significant changes on project cost and schedule. The SMEs assessed

Risk #	Status	Dependency	Project Phase	Summary Description Threat and/or Opportunity	Detailed Description of Risk Event (Specific, Measurable, Attributable, Relevant, Timebound) [SMART]	Risk Trigger	Type	Probability Correlation	Risk Impact ($M or Mo)	
(1)	(2)	(3)	(5)	(6)	(7)	(8)	(9)	(10)	[10a]	(11)
5	Active		Pre-construction	Threat	Three houses that were previously not identified as being National Register of Historic Places (NRHP) eligible were discovered within the corridor during a recent cultural resource survey update. These 3 houses create additional 4(f) impacts. This may trigger a Supplemental EIS. Noise and visual impacts may also trigger a Supplemental EIS, but this is less likely. If a Supplemental is needed the risk to schedule will be impacted most significantly, cost will be impacted, however not as significantly as schedule.	FHWA decision based on re-evaluation submittal	Cost	50%	MIN	0.10$M
									MAX	0.50$M
									Most Likely	0.30$M
				NEPA to EIS			0	Positive correlation	Master Duration Risk	
									MIN	12.0Mo
							Schedule		MAX	24.0Mo
				Threat					Most Likely	18.0Mo

Risk trigger: Submittal of NEPA Reevaluation documentation to FHWA. If it is determined the historic properties/4(f) impacts are significant, then the Supplemental will be necessary. This will impact shcedule due to the time necessary to prepare and process the Supplemental EIS.

The NRHP-eligible industrial facility became known recently (within last 2 months) and this has been discussed within department and with the federal agency. This risk can cause substantial delay for NEPA completion. This is mostly a schedule rather than cost risk.

Risk Matrix

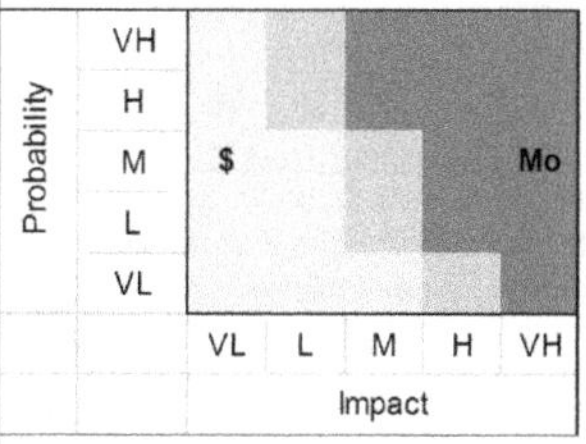

Figure 6.41 Section A—Risk 5

two risks (threat and opportunity) that may occur. The threat is called "Hazard materials" and the opportunity is called "Hazard materials 2." These two events are partial mutually exclusive relationships. Since the data about hazard materials is minimal, the SMEs concluded that *no clue events* is the best way to describe the probability of risks' occurrence. Figure 6.47 presents the hazard materials risks situation.

Figure 6.47 looks quite similar to Figure 4.9, doesn't it? It represents the classic no clue events of mutually exclusive risks.

Figure 6.48 presents two mutually exclusive risks where the events' probabilities of occurrences have better evaluations. The "NEPA to EIS" has about 50 percent probability of occurrence and "ENV. Permits" has 20 percent probability of occurrence only when the "NEPA to EIS" does not occur. In this case, the base alone will occur 40 percent of the time, as presented in Figure 6.48.

Risk #	Status	Dependency	Project Phase	Summary Description Threat and/or Opportunity	Detailed Description of Risk Event (Specific, Measurable, Attributable, Relevant, Timebound) [SMART]	Risk Trigger	Type	Probability Correlation	Risk Impact ($M or Mo)	
(1)	(2)	(3)	(5)	(6)	(7)	(8)	(9)	(10)	[10a]	(11)
6	Active	ME	Pre-construction	Threat	Will need a Shoreline Substantial Development Permit for work within 200' of the Black River. Timelines for review, public hearing, appeal period, and final approval can exceed a year.	Permit application submittal to the city officials	Cost	20%	MIN	0.10$M
									MAX	0.25$M
									Most Likely	0.15$M
				Env. permits			6			0
							Schedule		MIN	6.0Mo
									MAX	14.0Mo
				Threat					Most Likely	10.0Mo

Risk Matrix

This is an unknown entity in regard to timelines and complexity. It is premature to start discussions with regulatory agencies as design is too far out. This is mostly a schedule rather than cost risk.

Identify impacts and mitigation options early. Look for "in lieu fee" options and onsite mitigation potential. Early coordination with agencies and stakeholders will be critical; to obtain permit.

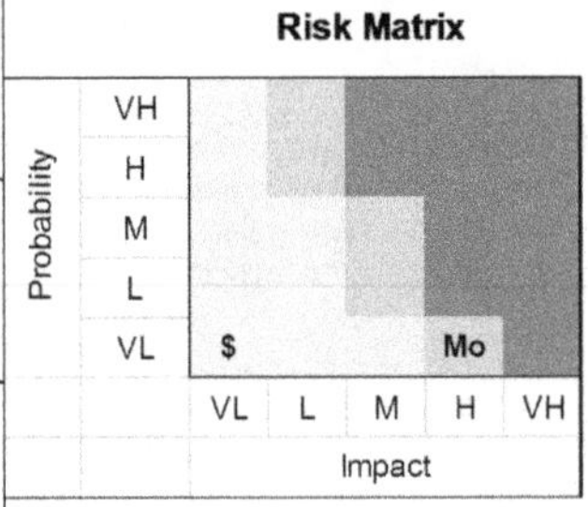

Figure 6.42 Section A—Risk 6

At this point in the process, users may run simulations and the model will provide the cost distribution for each phase of project delivery (PE, ROW, and CN) and the total project cost using the time of workshop prices (CY) or escalated values to the year of expenditure (YOE). We are not presenting premitigated results since the project went through a postmitigated risk analysis so we will present premitigated and postmitigated results on the same graph or in side-by-side tables.

The process of responding to risks may take two weeks to three months and requires in-depth analyses of each risk and measures that may be taken in order to optimize their effects. The project manager should avoid just reducing the probability of occurrence without good and documented justification. Any change on risk mesh must be documented and justifiable. At the same time, change of risks' impact values must be explained and documented.

Risk #	Status	Dependency	Project Phase	Summary Description Threat and/or Opportunity	Detailed Description of Risk Event (Specific, Measurable, Attributable, Relevant, Timebound) [SMART]	Risk Trigger	Type	Probability Correlation	Risk Impact ($M or Mo)	
(1)	(2)	(3)	(5)	(6)	(7)	(8)	(9)	(10)	[10a]	(11)
7	Active		Pre-construction	Threat		New data become available	Cost	10%	MIN	1.00$M
									MAX	4.00$M
									Most Likely	3.00$M
				Relocations	Design knowledge is low with regard to utilities at the time of this analysis. There is a risk that the estimated costs for utility relocations is low.		0		Master Duration Risk	
							Schedule		MIN	
									MAX	
									Most Likely	

fiber-optic vault (may not be able to relocate w/o extreme difficulty and $3-$4M)—this risk depends on the location of the vault—it may not impact the project at all—need to determine.

Risk Matrix

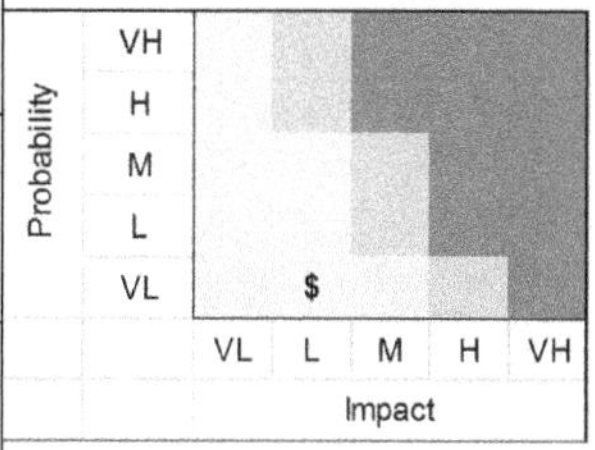

Figure 6.43   Section A—Risk 7

A common problem that risk managers face is the fact that if they just wait another week they will have better information and be able to provide a better response strategy. It is important to identify the response in a timely manner so that implementation of that response can begin as soon as possible.

The managers of section A have negotiated with the railroad owners and have reached an agreement that terminates Risk 2 (railroad) but it requires $10 M in extra funds (two additional RR crossings) that has to be included in the base. The agreement is a success since the schedule will be significantly improved and dollar-wise the cost is lower than the risk's expected value (about $12 M). The postmitigated scenario will run the model with Risk 2 retired.

The next risk that the project managers have mitigated is Risk 3. The project's site was surveyed for contaminated materials and the results came in favorable. The survey has found that the contaminated spots are not so numerous and the cleanup effort should be less expensive

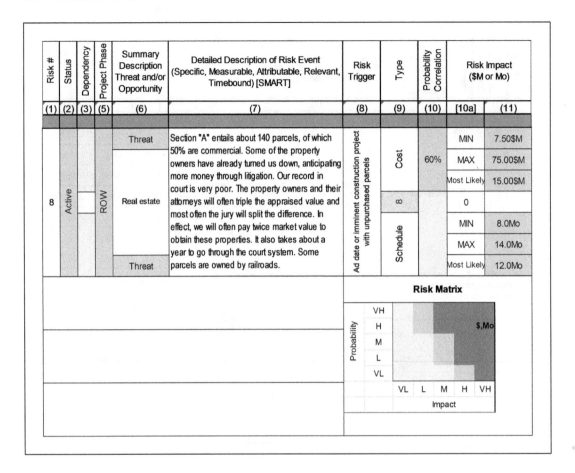

Figure 6.44   Section A—Risk 8

if the threat (Risk 3) occurs. The new revised description of postmitigated Risk 3 is presented in Figure 6.49.

Finally, Risk 9 (retaining wall) was terminated because the new data received from a noise level survey demonstrates that the noise impact will be minimal. That means that Risk 9 is retired on postmitigated analysis and Risk 10 (wall reduction) would have a 1-out-of-10 chances to occur.

The process of risk management continues as we specified in previous chapters but the changes introduced by these three mitigation actions have the managers very interested in learning about how much it may cost and how long it may take. So the postmitigated risk analysis was run with the adjustments presented in the previous paragraphs. The results are presented in Figures 6.50 to 6.55 following the same template. The graphs show results of premitigated and postmitigated scenarios including the base estimate.

Risk #	Status	Dependency	Project Phase	Summary Description Threat and/or Opportunity	Detailed Description of Risk Event (Specific, Measurable, Attributable, Relevant, Timebound) [SMART]	Risk Trigger	Type	Probability Correlation	Risk Impact ($M or Mo)	
(1)	(2)	(3)	(5)	(6)	(7)	(8)	(9)	(10)	[10a]	(11)
9	Active		Construction	Threat			Cost	25%	MIN	2.00$M
									MAX	25.00$M
									Most Likely	10.00$M
				Retaining wall	Retaining wall may need extending		0		Master Duration Risk	
							Schedule		MIN	
									MAX	
				Threat					Most Likely	

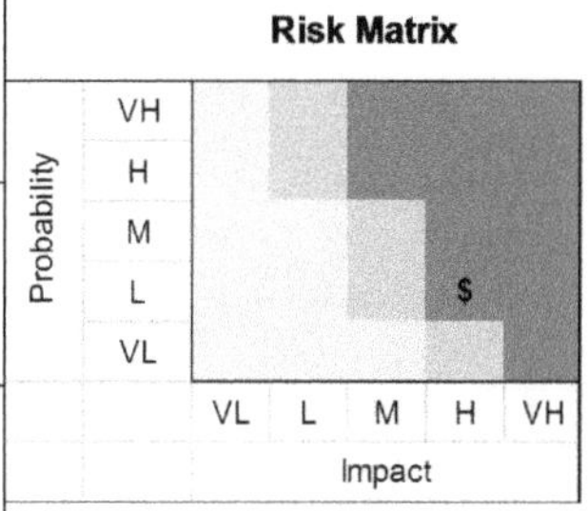

Risk Matrix

Probability: VH, H, M, L, VL — Impact: VL, L, M, H, VH — ($ marker at L probability / M impact)

Figure 6.45   Section A—Risk 9

The table from the right side shows the same results using numeric values. The base estimate (cost or end of activity) is presented at the top of the table. The CDF of the premitigated is represented by continuous line and the CDF of the postmitigated results is represented by a broken line. When cost is presented, it represents the estimated expenditure at delivery time (YOE).

Figure 6.50 shows the estimated total project cost, which ranges from about $288 M to $551 M when the postmitigated scenario is considered. The project experiences substantial cost and schedule reductions when the risk response plan is considered. The results provided at the postmitigated scenario may be used to define the project budget. The project managers must continue their efforts of risk response and manage the project to its budget.

Figure 6.50 shows that significant cost reduction occurs after the "railroad and retaining wall" risks were terminated and "hazard materials and city request" risks reduced because the survey

Risk #	Status	Dependency	Project Phase	Summary Description Threat and/or Opportunity	Detailed Description of Risk Event (Specific, Measurable, Attributable, Relevant, Timebound) [SMART]	Risk Trigger	Type	Probability Correlation	Risk Impact ($M or Mo)	
(1)	(2)	(3)	(5)	(6)	(7)	(8)	(9)	(10)	[10a]	(11)
10	Active	ME	Construction	Opportunity			Cost	10%	MIN	3.99$M
									MAX	4.01$M
									Most Likely	4.00$M
				Wall Reduction	Opportunity to save money on wall costs via slope modification.	New Design	10		0	
							Schedule		MIN	
									MAX	
									Most Likely	

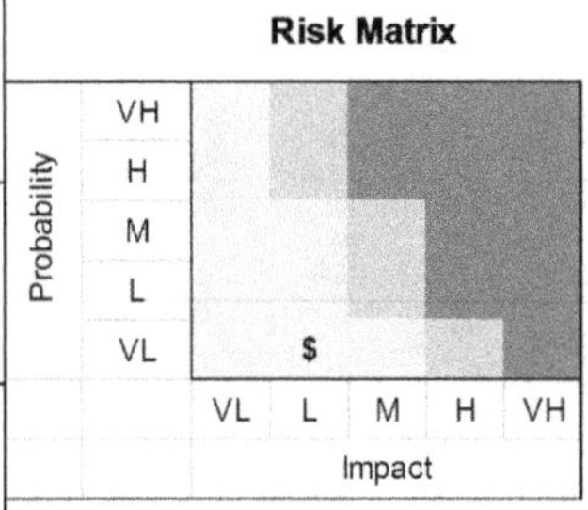

Risk Matrix

Figure 6.46   Section A—Risk 10

data demonstrated the hazardous materials are moderate and the city reduced its request. Despite the fact that base cost increases by about $10 M, the distribution of postmitigation is to the left of the premitigated distribution.

Figure 6.51 shows that the postmitigated scenario does not change the estimated cost of land. The estimated cost of land is significantly affected by "real estate" (Risk 8). At a 60 percent confidence level, the estimated cost is about 50 percent higher than the most likely base cost value. The risk must be the top priority of project management. The information available at the time of postmitigated analysis did not provide any support for reevaluation of this risk.

Figure 6.52 shows the estimated cost of preliminary engineering (PE). While the estimated base cost of PE does not change when the postmitigated scenario is considered, the cost distribution changes significantly mainly because the additional PE cost created by construction risks—two retired and two significantly reduced—was dramatically reduced. For example, the

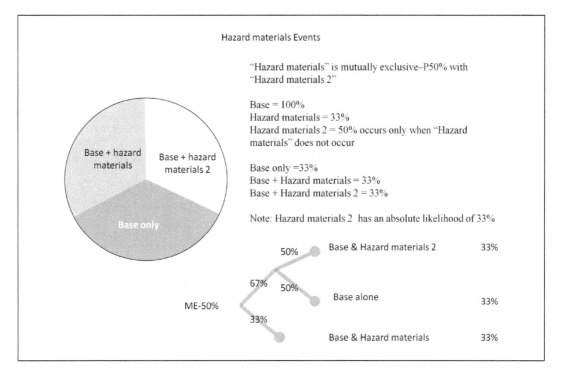

Figure 6.47 Section A—Hazard materials ME No Clue Treats

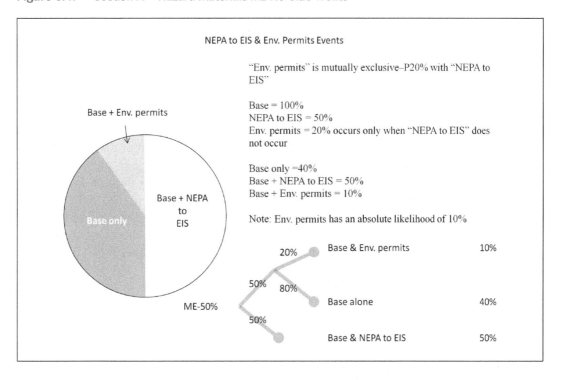

Figure 6.48 Section A—NEPA to EIS and Env. Permits ME–20 Percent

Risk #	Status	Dependency	Project Phase	Summary Description Threat and/or Opportunity	Detailed Description of Risk Event (Specific, Measurable, Attributable, Relevant, Timebound) [SMART]	Risk Trigger	Type	Probability Correlation	Risk Impact ($M or Mo)	
3	Active		Construction	Threat		Discovery of contaminated soils or petroleum storage tanks	Cost	10%	MIN	1.00$M
									MAX	10.00$M
									Most Likely	2.00$M
				Hazard Materials	Large parts of this project will be built in a railroad corridor and through an industrial site. Past and present hazardous materials generators along the alignment include gas stations, automotive repair shops, petroleum product storage sites and bulk plants, and an aluminum recycling facility.		0	Positive correlation	Master Duration Risk	
							Schedule		MIN	0.0Mo
									MAX	3.0Mo
				Threat					Most Likely	2.0Mo

Risk Trigger: Discovery of hazardous materials during the initial site assessment, and secondarily during construction.

Risk Matrix

Sites surveyed indicate that the contamination is less extensive and the cost and the delay for cleaning up is significantly lower.

Response strategies: Diligent research during the initial site assessment in order to avoid discoveries during construction. Avoid and minimize excavation in areas with known contamination.

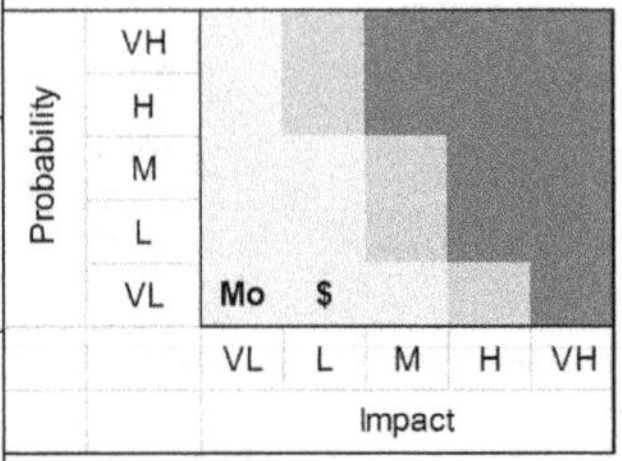

Figure 6.49 Section A—Risk 3 Is Postmitigated

negotiation with RR representatives that end up in termination of railroad risk causes reduction of construction cost and reduction of PE cost.

Figure 6.53 shows the construction cost estimate distribution. The effect of market condition on the postmitigated scenario is apparent. The hump at the histogram right side indicates the worse than planned market conditions while the semi-hump at the histogram left side indicates the better than planned market conditions. The postmitigated market condition is the same as the premitigated market condition but they show off differently on graphs. This happens because the postmitigated scenario has a simpler risk mesh than the premitigated scenario.

The advertisement date is presented in Figure 6.54. As expected, the distribution of the premitigated and postmitigated scenarios are similar and they reflect the effect of two major schedule risks. The histogram shows three humps, which are the results of three major events. We may reasonably assume that: (1) the hump from the left side is given by baseline schedule alone; (2) the middle hump comprises base values plus "real estate and env. permits" (Risks 8

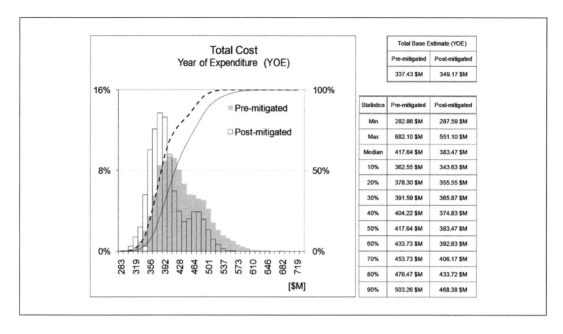

Figure 6.50   Section A—Estimated Total Cost

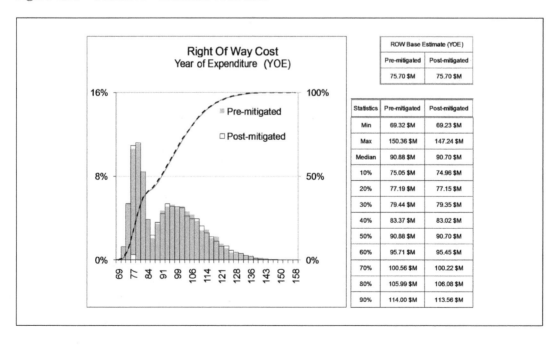

Figure 6.51   Section A—Estimated Cost of Land

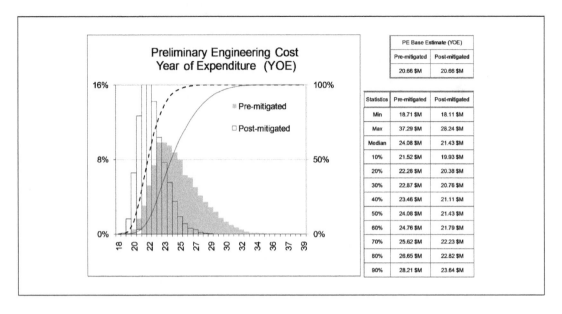

Figure 6.52 Section A—Estimated Preliminary Engineering Cost

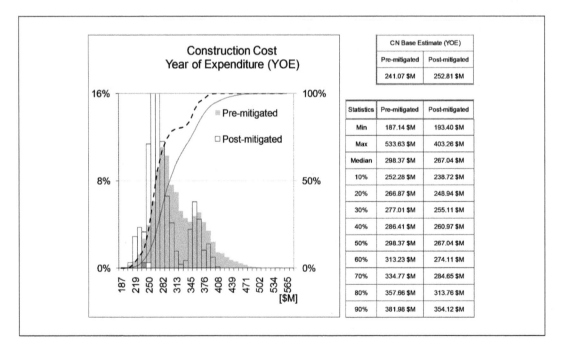

Figure 6.53 Section A—Estimated Construction Cost

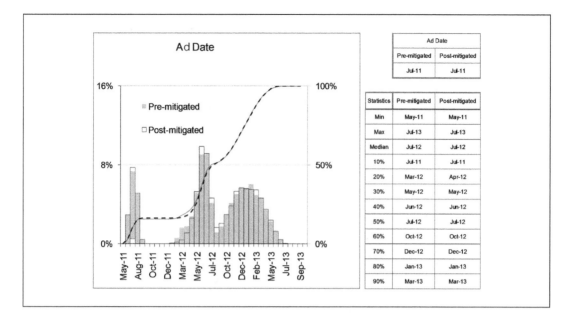

Figure 6.54   Section A—Estimated Advertisement Date

and 6); and (3) the hump from the right side is given by base values and "NEPA to EIS" (Risk 5).

Figure 6.55 shows the distribution of end of construction date. It does not have any particular shape since the distribution includes the variance of ad date and variability of construction duration plus three major construction schedule delays. The postmitigated scenario creates significant schedule reduction (about one year) to the project.

The tornado diagrams and risks map of the premitigated scenario have been presented in Chapter 3 (Figures 3.40 and 3.42, respectively).

The postmitigated scenario has a different outcome since four risks that affect both cost and schedule have been altered to reflect the postmitigated scenario. The tornado diagram of the postmitigated scenario is shown in Figure 6.56 and the new risks map is presented in Figure 6.57.

The "real estate" is maintaining its top position on the cost diagram and second position on the schedule diagram. The "railroad and retaining wall" risks are no longer on the figure since they were retired. "Hazard materials" risk moved two places down on the cost area and is the last risk on the schedule area.

The risks map indicates that "real estate" risk constitutes first priority because of its two components (cost and schedule). "City request and NEPA to EIS" risks are very high priority because of their schedule component. For this project, accordance appears between the tornado diagrams and risks map recommendations since both convey the same message. The process

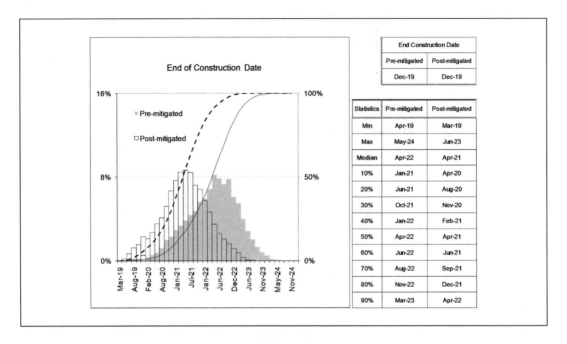

Figure 6.55 Section A—Estimated End of Construction Date

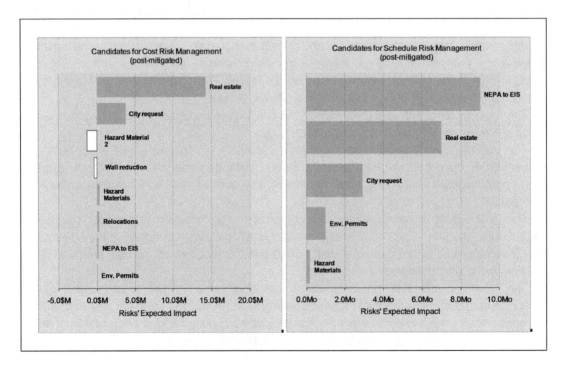

Figure 6.56 Section A—Candidates for Risk Response

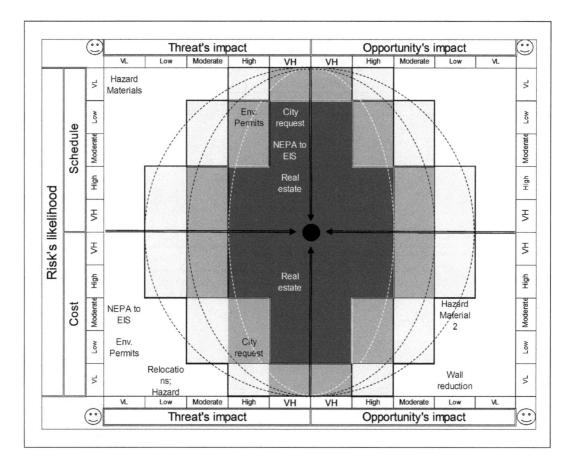

Figure 6.57   Section A—Risks Map

of risk assessment have to be continued and the model has to be rerun when any significant changes occurs.

Case Study No. 2: Eastside Corridor — Corridor Level

The Eastside Corridor—Corridor Level may be applied to premitigated or postmitigated section A. The project owner wanted to have an estimate immediately after the workshop. In this case, the Eastside Corridor—Corridor Level (EC-CL) was analyzed using premitigated section A base cost and risks. The analysis algorithm is described in Chapter 3, section "Program Level Analysis."

The real estate SMEs decided to deviate from the algorithm presented in Chapter 3 by combining section B and section C in one activity, and for each land acquisition activity they defined a base cost with uncertainty and a risk associated to it as presented in Figures 6.58 and 6.59.

Risk #	Status	Dependency	Project Phase	Summary Description Threat and/or Opportunity	Detailed Description of Risk Event (Specific, Measurable, Attributable, Relevant, Timebound) [SMART]	Risk Trigger	Type	Probability Correlation	Risk Impact ($M or Mo)	
(1)	(2)	(3)	(5)	(6)	(7)	(8)	(9)	(10)	[10a]	(11)
17	Active		ROW	Threat			Cost	100%	MIN	62.08$M
									MAX	102.26$M
									Most Likely	73.04$M
				ROW-Section "D" Base	Base cost 73$M with a MIN=−15% and MAX=+40% of it; Base Duration 12Mo with a MIN=−15% and MAX=+40% of it.		0		Master Duration Risk	
							Schedule		MIN	10.0Mo
									MAX	17.0Mo
				Threat					Most Likely	12.0Mo

This uncertainty includes base variability, market conditions, minor risks.

Risk Matrix

(1)	(2)	(3)	(5)	(6)	(7)	(8)	(9)	(10)	[10a]	(11)
18	Dormant		ROW	Threat		Construction schedule	Cost	50%	MIN	5.00$M
					To date eminent domain is not being used. However the 13 commercial parcels in this section will be more difficult to acquire, and there is a strong likelihood that condemnation or extensive administrative settlements may be required. Should either occur, both the cost and schedule will be impacted.				MAX	31.00$M
				ROW-Section "D" Risks					Most Likely	27.00$M
							18		0	
		Positive					Schedule	Positive correlation	MIN	12.0Mo
									MAX	24.0Mo
				Threat					Most Likely	18.0Mo

Risk Trigger details: If we are not able to acquire all the parcels, the ad date or at least the award date will be extended. Most commercial owners in this area are relatively sophisticated business people, so there is a reasonable probability that condemnation or administrative settlements will occur.

Upper management and the Project Engineer's office are well aware of this potential risk. This is considered a 50% risk factor that needs to be recognized and dealt with.

We will continue to acquire the properties in this area under the voluntary seller approach, but if condemnation is required, those parcels should be turned over to the AG's office about 18 months in advance of the ad date.

Risk Matrix

Figure 6.58 EC-CL: Section D, Land Acquisition

Risk #	Status	Dependency	Project Phase	Summary Description Threat and/or Opportunity	Detailed Description of Risk Event (Specific, Measurable, Attributable, Relevant, Timebound) [SMART]	Risk Trigger	Type	Probability Correlation	Risk Impact ($M or Mo)	
(1)	(2)	(3)	(5)	(6)	(7)	(8)	(9)	(10)	[10a]	(11)
19	Dormant		ROW	Threat			Cost	100%	MIN	81.36$M
					Base cost 90.4 $M with a MIN = -10% and MAX = +30% of it				MAX	117.52$M
				ROW-Section "B & C" Base					Most Likely	90.40$M
					Base Duration 70 Mo with a MIN = -10% and MAX = +30% of it.		Schedule	0	Master Duration Risk	
								Positive correlation	MIN	63.0Mo
									MAX	91.0Mo
				Threat					Most Likely	70.0Mo

This uncertainty includes base variability, market conditions, minor risks.

Risk Matrix

(1)	(2)	(3)	(5)	(6)	(7)	(8)	(9)	(10)	[10a]	(11)
20	Dormant	Positive	ROW	Threat		Ad date or imminent construction project with unpurchased parcels	Cost	60%	MIN	0.50$M
									MAX	90.00$M
				ROW-Section "B & C" Risk					Most Likely	30.00$M
							Schedule	20	0	
								Positive correlation	MIN	12.0Mo
									MAX	24.0Mo
				Threat					Most Likely	18.0Mo

Risk Trigger details: If we are not able to acquire all the parcels, the ad date or at least the award date will be extended. Most commercial owners in this area are relatively sophisticated business people, so there is a reasonable probability that condemnation or administrative settlements will occur.

Upper management and the Project Engineer's office are well aware of this potential risk. This is considered a 60% risk factor that needs to be recognized and dealt with.

We will continue to acquire the properties in this area under the voluntary seller approach, but if condemnation is required, those parcels should be turned over to the AG's office about 18 months in advance of the ad date.

Risk Matrix

Figure 6.59   EC-CL: Sections B and C, Land Acquisition

Risk #	Status	Dependency	Project Phase	Summary Description Threat and/or Opportunity	Detailed Description of Risk Event (Specific, Measurable, Attributable, Relevant, Timebound) [SMART]	Risk Trigger	Type	Probability Correlation	Risk Impact ($M or Mo)	
(1)	(2)	(3)	(5)	(6)	(7)	(8)	(9)	(10)	[10a]	(11)
21	Dormant		Construction	Threat			Cost	100%	MIN	176.55$M
					The impact represents the Section 4 uncertainty which includes base variability, market condition and risk events. It is assumed that the validated cost coincides with the Most Likely value, the MIN is equaled to 0.8 of the Most Likely value and MAX is equaled to 1.5 of the Most Likely value.					MAX	331.04$M
				CN - Section "B" Base					Most Likely	220.69$M
							0		Master Duration Risk	
							Schedule	Positive correlation	MIN	28.00$M
									MAX	52.50$M
				Threat					Most Likely	35.0Mo
(1)	(2)	(3)	(5)	(6)	(7)	(8)	(9)	(10)	[10a]	(11)
22	Dormant	Positive	Construction	Threat			Cost	100%	MIN	231.04$M
					The impact represents the Section 5 uncertainty which includes base variability, market condition and risk events. It is assumed that the validated cost coincides with the Most Likely value, the MIN is equaled to 0.75 of the Most Likely value and MAX is equaled to 1.7 of the Most Likely value.					MAX	523.69$M
				CN-Section "C" Base					Most Likely	308.05$M
							22		0	
							Schedule	Positive correlation	MIN	26.0Mo
									MAX	60.0Mo
				Threat					Most Likely	35.0Mo
(1)	(2)	(3)	(5)	(6)	(7)	(8)	(9)	(10)	[10a]	(11)
23	Dormant	Positive	Construction	Threat			Cost	100%	MIN	155.97$M
					The impact represents the Section 6 uncertainty which includes base variability, market condition and risk events. It is assumed that the validated cost coincides with the Most Likely value, the MIN is equaled to 0.70 of the Most Likely value and MAX is equaled to 2.0 of the Most Likely value.					MAX	445.62$M
				CN-Section "D" Base					Most Likely	222.81$M
							0		Master Duration Risk	
							Schedule		MIN	32.0Mo
									MAX	92.0Mo
				Threat					Most Likely	46.0Mo

Figure 6.60   EC-CL: Sections B, C, and D Construction

Each figure displays two components: (1) base variability and (2) risk. It was assumed that risk is positive correlated with the corresponding base. The schedules presented for the base and risk are for information only, since the corridor level schedule is driven by the section A end of construction date and construction duration of each of sections B, C, and D, considered in series.

Based on the algorithm presented in Chapter 3, the construction estimate is defined for each section as base cost and duration. The durations are in series so the end of the program is defined by the summation of "End CN" section A plus sections B, C, and D duration added together. The experts have chosen the scenario of having cost of each uncertainty positive correlated. Figure 6.60 presents the uncertainties of construction cost and schedule of sections B, C, and D.

Modeling of the EC-CL requires advanced understanding of how the model works (recoding) and is not recommended to be done by regular users. Once the model is set up, regular users may use it and change values to fit new data.

The owners were interested in total program cost and end of construction program. For this scale of analysis (quality of data input) their request made a lot of sense. Figure 6.61 shows the estimated total program cost in year of expenditure and Figure 6.62 presents the end of program construction.

The total estimated values of the Eastside Corridor (Figure 6.61) encompasses a wide range from about $1.5 B to $2.6 B, and is distributed in a shape that resembles normal distribution. The majority of data were expressed through uncertainty (100 percent probability of occurrence).

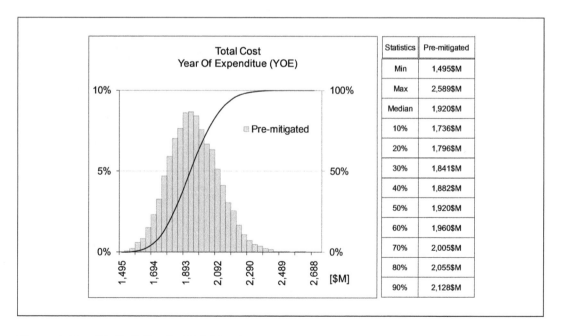

Statistics	Pre-mitigated
Min	1,495$M
Max	2,589$M
Median	1,920$M
10%	1,736$M
20%	1,796$M
30%	1,841$M
40%	1,882$M
50%	1,920$M
60%	1,960$M
70%	2,005$M
80%	2,055$M
90%	2,128$M

Figure 6.61   EC-CL: Estimated Total Cost

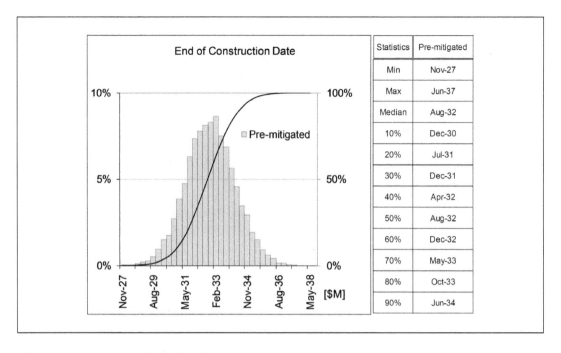

Figure 6.62   EC-CL: Estimated End of Construction Date

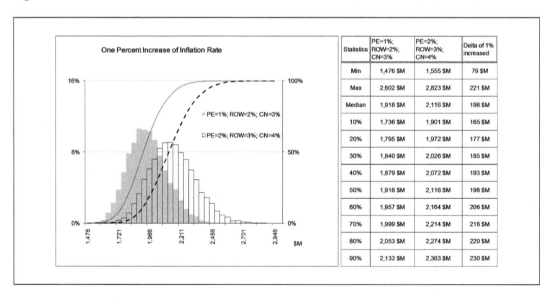

Figure 6.63   EC-CL: Estimated Cost Increase Produced by 1 Percent Increase of Inflation Rate

The program may be accomplished as soon as the end of 2027 and as late as mid-2037. It is approximately 10 years' range for the end of program construction (see Figure 6.62).

Figures 6.61 and 6.62 show results that reflect the experts' knowledge at the time of the estimate. It is highly expected that even these wide ranges will not hold because the data (program conditions and inflation rate) may change. The Eastside Corridor has a long delivery schedule, which makes the cost very sensitive to variation of inflation rate.

Figure 6.63 shows the effect on total program cost induced by only 1 percent increase of inflation rate. Each of the program phases would have an increase of its inflation rate by 1 percent. The effect is significant. Figure 6.63 indicates an average increase of approximately $200 M. Two hundred million dollars is a lot of money but we are estimating a very long program. It is similar to having a mortgage when we pay twice or three times the initial cost. Knowing that the project is so sensitive to inflation, the project owners may plan ahead and inform their stakeholders about what may happen and why the schedule is an important cost factor. Time is money!

SUMMARY

This chapter has presented the RBES, which is a tool that the readers may use at their discretion. The simplified version of RBES is presented and may be downloaded at www.cretugroup.com.

We recommend that any RBES user first attempt to model the "micro-project" and Eastside Corridor (section A) projects before using RBES in an actual application. This will assist with developing some basic understanding of the model and what effects the different data input have on the outcomes.

A more thoughtful approach would be to attend training classes for RBE and RBES, and follow that with participating or observing workshops conducted by other professionals. After that you will be more comfortable using the RBES on a real project and be able to identify where additional coaching could be helpful.

Some users have found risk conditionality confusing and have tried to stay away from it. It may be a good temporary strategy of gradually learning, but the reader must understand that the capturing of risk conditionality is critical when it exists. Practice using simple situations (just two risks) and using the "common sense test" will help improve your understanding of risk analysis and build confidence on your own skills.

Warning: Do not rush and do not think you are an expert until you are an expert. Ask yourself questions; use the common sense test; and, if you cannot find a satisfactory answer, ask for guidance.

ENDNOTES

1. O. Cretu and T. Berends, "Risk-Based Estimate Self-Modeling," 2009 AACE International TRANSACTIONS, Seattle, WA, June 28–July 1, 2009; O. Cretu, "Risk-Based Estimate of Transportation Infrastructures," The International PIARC Seminar—Managing Operational Risks on Roads, November 5–7, 2009, Iasi, Romania.

2. O. Cretu, T. Berends, and R. Stewart, "Reflections about Base Cost Uncertainty," Society for Risk Analysis Annual Meeting 2009, Risk Analysis: The Evolution of a Science, December 6–9, 2009, Baltimore, Maryland.

3. R.L. Iman, J.M. Davenport, and D.K. Zeigler, "Latin Hypercube Sampling (A Program User's Guide)": Technical Report SAND79–1473, Sandia Laboratories, Albuquerque (1980).

4. Risk-Based Self-Modeling Spreadsheet, www.cretugroup.com.

5. O. Cretu, T. Berends, R. Stewart, and V. Cretu, Risk-Based Estimate—Keep It Simple... International Mechanical Engineering Congress and Exposition, November 13–19, 2010, Vancouver, BC.

CHAPTER 7

RISK-BASED ESTIMATE
WORKSHOP

INTRODUCTION

The risk-based estimate workshop, or cost risk analysis workshop, identifies and quantifies risk events and uncertainties that have a consequential impact on the project. The RBE workshop, hereafter referred to as the *workshop*, finalizes data collection and brings to a closure any disagreements that may occur during either base cost review (see Chapter 3) or risk elicitation (see Chapter 4).

This chapter shows a practical application of the RBE process as described in Chapters 3 and 4 by demonstrating its general implementation. Actual implementation may take different forms and address different levels of complexity. Our aim is to outline both a generic and reasonable approach, and to demonstrate the RBE process as we practice it.

The most important recommendation that we can make that is critical for any workshop is to "watch for workshop integrity." If the workshop integrity is in doubt then the entire effort is worthless and the RBE process loses credibility. We consider that workshop integrity may be affected when outside or inside elements influence the data.

The *process* of the workshop (described later in this chapter) can be easily understood and executed; however, from our experience, the most important elements are not those related to the process technically, but rather the *atmosphere* of the workshop. So, first, a few words on that.

Workshop leads need to recognize and eliminate the influences coming from either RBE participants or outside forces. The only way to have data that is as accurate as our knowledge is to create an environment where estimators and subject matter experts can freely participate in the workshop. As was previously stated in Chapter 2, *estimators should be shielded from pressures to keep estimates within programmed or desired amounts based on funding availability. Estimators should be free to establish what they consider to be a reasonable estimate based on the scope and schedule of the project and the bidding conditions that are anticipated.* In conclusion, participants should be both unfettered (free to speak their minds) and accountable (liable for their inputs).

The following are a few observations we found common to successful workshops:

- Executive-level decision makers rely on workshop results to make their decisions.
- Workshop participants received preworkshop training. This improved with conducting advance risk interviews and overall communication among participants.
- The KISS mantra is followed.
- Results are analyzed and tracked at a set interval.

Workshop Tangible Outcomes

The workshop provides data to the risk modeler for further analysis in the form of two major items:

1. Base estimate
   - validated assumptions
   - validated base cost and schedule as deterministic number
   - uncertainty associated with base cost
   - variability of base schedule
   - risks' markups
   - inflation rate (This is decided by a small group of professionals familiar with the regional and national markets; the workshop participants should not be engaged in debating the inflation since it is going to be a useless and endless task.)

2. A completed list of risks identified and quantified during the entire process. This list contains all the information related to risks:
   - project's constraints
   - project flowchart
   - detailed narrative risk description—(SMART approach must be employed.)
   - probability of occurrence
   - impacts—cost and schedule
   - risks' conditionality—dependency and correlation
   - possible mitigation strategies for identified risks

Note: All the information provided by the workshop should be written and detailed at a level that is both efficient and relevant. The analysis must use only documented information, and the information must be clear. Just because the modeler (individual or individuals who create the simulation model) participated in the workshop does not provide grounds for not having documentation of any actions included on the model.

Workshop Steps

The workshop finalizes, validates, and documents its tangible products through a process of informing, debating, and consensus. There are several steps that the workshop must go through in order to deliver tangible outcomes. Each step brings value to the workshop and it is important that the workshop spend a balanced time discussing each of them.

When one step encounters difficulty or experiences more contentious issues, the process must not stall and the workshop coordinator must intervene to expedite the process. Depending on the issue of concern, there are different methods to expedite the process, but the single most efficient way of avoiding difficult issues during the workshop discussions is ensuring proper *preparation* for the workshop. The preworkshop activities are designed to remove the controversy around different topics and develop consensus ahead of the workshop.

The preworkshop activities relate to the workshop in the same way that risk management relates to the project. One of the objectives of the risk management function is to avoid surprises. Preworkshop activities must eliminate or assuage all major disagreements. The workshop is not the place for resolving conflicts, but rather, it is the place to announce resolutions and validate them.

If the preworkshop activities are successfully conducted, the workshop and postworkshop steps presented next will likely go smoothly and be effective.

- Review or validate the basis of the estimate by the lead cost and schedule reviewer and by subject matter experts from inside and outside of the organization.
- Document assumptions and constraints used in developing the estimated project cost and schedule range.
  a. Assumptions must be validated. Those that are not validated must be evaluated as risks.
  b. Constraints must be included on risk analysis algorithm.
- Replace the explicit and hidden project contingencies with key identifiable risks that can be more clearly understood and managed. It may be acceptable to include in the base cost a small percentage of the deterministic estimated cost as a contingency available to the project manager.
- Under the direction of the risk lead, identify and quantify the project's key events that can cause significant deviation from the base cost and/or schedule. Usually, risks are already

identified and quantified through advanced risk elicitation meetings and the workshop validates them; if necessary, risks may be changed.

Since the workshop participants examine the entire risk mesh, now is the time for clarifying risks conditionality. The risk leads are responsible for making sure that all participants understand risks' impact and the relationships among them.

- Discuss and develop concepts for responses to risks that could impact the cost and/or the schedule of the project. Provide the project team with actionable information on risk events that allow them to manage risks.

PREWORKSHOP ACTIVITIES

Preparation for the workshop may take one or more meetings depending on the project size, complexity, and knowledge of the participants. The project manager should work with the workshop coordinator and cost-risk team to identify the best combination of participants for each meeting. The goal is to effectively use time for all parties in a manner that ensures a sound and objective analysis.

The criterion for project workshop participation has to be: "Who is the most knowledgeable to identify and clarify issues that may or may not occur?" The participation criterion should include not only professionals with technical expertise but also professionals with problem-solving and team-building skills since the workshop participants should (1) be involved and (2) be heard—in relation to their responsibility and/or expertise.

Advanced preparation for the workshop should be the focus of the risk lead, cost lead, and project manager because they will be tasked with helping to develop the project flowchart, assemble project cost and duration estimates, and develop a list of risks that could have a significant impact on the project's schedule and/or cost.

The initial meeting (prep session) will identify who should participate in the upcoming workshop and advanced risk elicitation interviews. During the prep session the cost lead learns about the project's basis of estimate and has initial discussions about the existing estimate for the cost and schedule. At this time the project flowchart is developed and the duration of each flowchart activity is penciled down.

The main function of the prep session is to initiate the process and assign roles and responsibilities to the RBE team members. It is a little more than "team alignment," as it is commonly known within the normal project management process. Table 7.1 presents the recommended key players in the project workshop.

After the prep session and before the workshop, the cost lead and assigned SMEs review the project base cost and schedule estimate and provide recommendations for changes, if any. The new cost and schedule estimate proposed by the base cost team should be reviewed and agreed upon by affected project team disciplines prior to the workshop.

The agreement on base cost prior to the workshop is important for developing a high quality estimate and for enhancing the credibility of the process. Nothing can be more damaging to the credibility of the process than discovering estimating errors during the workshop or having major disagreements at the last moment.

TABLE 7.1   Workshop Team Participants

Project Team	Roles & Responsibilities
*Project Manager	Project resource
*Estimator	Prepare and document project estimate
*Scheduler	Prepare and document project schedule
*Lead Designer	Primary resource for design questions
Key Technical Experts	Specialty groups as needed
Subject Matter Experts (SMEs)	**Roles & Responsibilities**
Project Team Experts Agency Experts Other Stakeholders External Consultants	Internal SMEs work with external SMEs to review and validate project cost and schedule estimates. They provide objective review and comments regarding project issues, risks and uncertainty. At the end of the workshop the SMEs should provide a brief summary of their thoughts about the workshop
Cost-Risk Team	**Roles & Responsibilities**
*Risk Lead	Conducts risk elicitation and manages meeting during risk elicitation
Risk Lead Assistant	Assists with risk elicitation and meeting management during risk elicitation
*Cost Lead	Conducts base cost and schedule review and validation; manages the meeting during the review
Cost Lead Assistant	If needed, assists the cost lead position, as appropriate
Workshop Coordinator	Coordinates the agenda and participants' discussions, works with the project manager to ensure the success of the workshop

* These participants should also attend the prep session.

Prior to the workshop, the risk lead should meet with specialty groups and elicit risks (both threats and opportunities) that have significant effects on project objectives. The methods of elicitation are described in Chapter 4. The risk elicitation may take any form as long as the process is effective and efficient.

The focus is on significant risks but other risks may be captured while recognizing that they may not be included in the analysis. The goal of the advanced risk interviews is to completely identify and quantify (full description) all significant events that may affect the project triad. We recommend that each risk identified be accompanied by suggestions on how to respond to it.

The preworkshop base cost estimate, flowchart, and the advanced risk interviews findings should be submitted to the workshop coordinator prior to the meeting in order to assess the readiness for the workshop. If significant dissonance exists, the workshop should be postponed until the agreements are in place, otherwise it is possible to ruin the entire effort. The decision to postpone a workshop is difficult to make; however, sometimes it must be made and when is made must be supported by the project decision makers.

The best workshops, in terms of effectiveness and efficiency, are those that have conducted ample advance work, particularly in the areas presented in previous paragraphs.

CONDITIONING WORKSHOP PARTICIPANTS

Prior to the actual workshop, participants need to know what to expect and what is expected of them. Participants are advised to avoid bias and to be impartial during the discussions that ensue at the workshop. Individuals need to listen to all opinions and not stubbornly advocate a predetermined point of view. Chapter 4 describes procedures for conditioning the project team and subject matter experts in order to prepare them for the workshop. The risk and cost leads are expected to be aware of potential biases as they conduct their respective portions of the workshop.

Project teams, particularly early in project development, are often optimistic about their project when the estimate is generally at the low end. Optimism bias has been observed to reverse itself as a project approaches completion of design when the project managers may become increasingly guarded about the financial needs of the project and pressure the estimators for high values. The discussion on prospect theory presented in Chapter 5 covered this in depth. It may be a significant shift from optimism to pessimism when an estimate is intentionally moved from low values to high values in order to make sure there is enough money for the project. If possible, a short training class is recommended to all participants involved in cost risk analysis, and this training may be conducted by the workshop coordinator or risk lead.

There may be situations where a significant difference of opinion has arisen among workshop participants. Usually the ability to capture inputs in ranges meets the needs of participants offering input. For example, if one participant states, "This risk event could cause $2.5 million in additional cost . . ." and another says, "This risk event could cause up to $10 million in additional cost . . ." we can simply offer to capture the risk with a $2.5 to $10 million impact range—typically, this will satisfy the parties with differing opinions about the impact. (Note: Persons offering opinions must be able to state why they have the opinion and document information used to develop the opinion.)

If the previous example has a base cost variability of approximately $3 million, then a better approach may exist. The parties may agree on identifying specific conditions for each event and then elicit two different risks while paying close attention to their conditionality. There is no "silver bullet" solution; however, there may be a number of good compromises that can be made to alleviate discontent (see Chapter 4, the section titled "The Solution").

In other cases it may be appropriate to evaluate additional scenarios that address the different opinions being offered. This is practical only if major decisions are made in advance of the workshop that addresses it. The scenario analysis is a procedure that we recommend to be used by the project team during their efforts of managing risks.

If a strong difference of opinion persists and the options previously presented do not resolve the matter, it is recommended that the necessary data be gathered and that the relevant parties meet to review and discuss the matter outside of the workshop setting.

Strive to use objective data, with guidance from the risk lead and cost lead, to reach an agreed-upon input. If, after a concerted effort to reach a consensus decision, disagreement still exists, it may be necessary to adopt a solution and document the dissenting opinion in the report.

Data objectivity when applied on RBE may have a wide range. What one professional considers very objective, another one may consider purely subjective. It depends on each individual's professional experience and, in many cases, it may be difficult to discern which one is better. The following list provides some guidance regarding data reliability:

- Scientific studies provide more reliable data than case studies.
- Many observations indicate more reliable conclusions than the conclusions drawn from one of few observations.
- Published data implies better information than unpublished data.
- Similar projects provide better data than unique projects.

WORKSHOP OBJECTIVES

The overall workshop objectives are:

1. Validate the project's basis of estimate, assumptions, constraints, and project flowchart. Advance preparation is important since any surprise brought up during the workshop may derail the activities.

2. Validate the base cost and schedule estimates. Advance base cost and schedule review and their acceptance are crucial for the quality of the estimate. The base estimate is the component that provides the center of the estimate range. Any errors in the base estimate translate in linear form to the published estimate.

3. Identify and quantify the uncertainties and risks. Advanced risk assessment elicitation increases the quality of the process and provides better understanding of risks and their conditionality. Risks and uncertainties define the edges of the estimate and risk analysis and management focuses on the edges.

RISK IDENTIFICATION PROCESS

Risk Categories

This item warrants additional discussion. The *PMBOK* discusses the use of a risk breakdown structure (RBS). An RBS is essentially a modified version of a work breakdown structure (WBS) where the aggregation describes risks instead of activities. This is a useful approach to identifying risk categories; however, the authors have prepared a simplified version of this that is tailored to construction projects. There is no need to reinvent the wheel each time, and the RBS shown in Figure 7.1 should serve well for most projects with minimal modification.

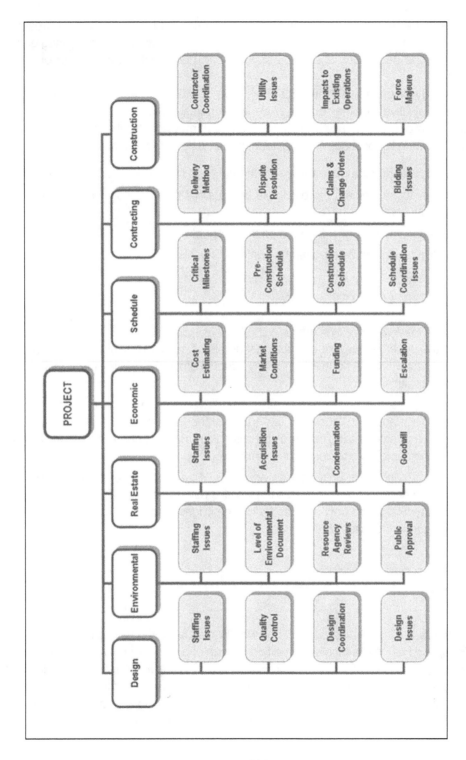

Figure 7.1 Example of a Risk Breakdown Structure

Who Identifies the Risks?

It is impossible to analyze risks unless we have first identified them. This seems pretty obvious, yet it seems that many of the projects the authors have encountered failed to do this. This is usually a symptom of ignorance, overconfidence, a lack of organization, or some combination thereof. The fact of the matter is that every project will face unique risks, no matter how simple and straightforward it may appear.

The identification of project risks should occur within the context of a facilitated, multidiscipline team effort. The organizer of the event, most often the project manager, should identify a specific time and place to identify project risks. Typically, one to two hours is sufficient for most projects; however, very large or complex projects may take additional time.

Participants on the risk management team should be selected to represent the various technical disciplines involved and may also include stakeholder representatives if appropriate. It is not necessary that all of the participants be a part of the project team; in fact, it is useful to include people who are not. Senior, more experienced participants are desirable, as much of the success of the risk management process is dependent upon the quality of the input. This input is as likely to be based on experiential knowledge as it will be on practical or technical knowledge.

The Risk Identification Process

The identification of project risks is the second step of the risk management effort. Another meeting should be planned where the risk management team can get together to discuss the unique risks that the project may face at some point. Generally, a one- to two-hour facilitated meeting is sufficient to achieve this. The focus of the meeting should be structured around fleshing out the risk categories that were identified in the risk planning meeting, as was discussed in the previous chapter.

Personally, I like to hold a team brainstorming session where I will stand by a flipchart and write down potential risks by category as the team throws them out. Another way to do this is by use of a personal computer and a multimedia projector. Either way is fine so long as you record the risks as they come. It is always best to do this in a way that everyone can see them; that way you can be sure that concepts are being recorded properly. It also stimulates thinking by allowing others to build on previous risk concepts. Tape the completed sheets on the wall and keep going! The idea is quantity. Try to discourage editing or filtering at this point—that will occur later.

Using an risk breakdown structure here will really make this much easier—it is akin to following a roadmap or checklist. If you have a predefined list of risks, be sure to run through them as well to make sure that all of the bases have been covered.

The best resource for generating risks, however, will be the experience and knowledge of the team. As the risk lead or facilitator, it is useful to ask the group rhetorical questions to get them engaged. Here are a few examples:

- Did we run into geotechnical problems at this location when the original foundations were installed?
- Does the local contracting community have the experience to construct this type of bridge? Will this affect bid prices?

- Are we at risk of getting sued for seriously disrupting access to the shopping mall next door during construction?

- Is 18 months enough time to finish design? Do we have enough resources to meet the *notice to proceed* milestone on the current schedule?

- How firm is our assumption that we can obtain the materials in the vicinity of the project? What if we can't?

This type of facilitation technique is referred to as Socratic questioning. Rhetorical questions are used to elicit responses. I assure you that if you ask the group these kinds of questions, you will get a response. This is a great way to lead as it is both unassuming yet effective at the same time.

Other possible ways of generating risks aside from team brainstorming or checklists is to use a technique called SWOT analysis. SWOT stands for strengths, weaknesses, opportunities, and threats. These four factors are defined as follows:

- **Strengths**—Attributes of the project that are helpful to achieving the objective

- **Weaknesses**—Attributes of the project that are harmful to achieving the objective

- **Opportunities**—External conditions that are helpful to achieving the objective

- **Threats**—External conditions that are harmful to achieving the objective

An example of SWOT analysis is shown in Figure 7.2. SWOT analysis is a useful tool for framing questions and eliciting information from the group. There are various software programs available for download on the Internet that can create more elaborate diagrams with added features. However, the real interest here is in using this as an aid in identifying risks.

The Risk Register

Once the risks for the project have been identified they should be organized into tabular format and additional information added to provide further detail. Most risk registers include the following information:

- **Reference number**—An identification number is commonly assigned to each risk in order to provide a convenient reference.

- **Risk category**—Indicate which category the risk belongs to. This should relate to the risk breakdown structure prepared earlier.

- **Risk type**—Indicate whether the risk is a threat or an opportunity.

- **Title**—Label the risk with a short, descriptive phrase that summarizes the risk.

- **Risk description**—A concise description of the risk event. It is recommended that the description adhere to the SMART protocol. SMART is an acronym that is short for specific, measurable, attributable, relevant, and time bound.

- **Trigger**—What event will cause the risk to occur?

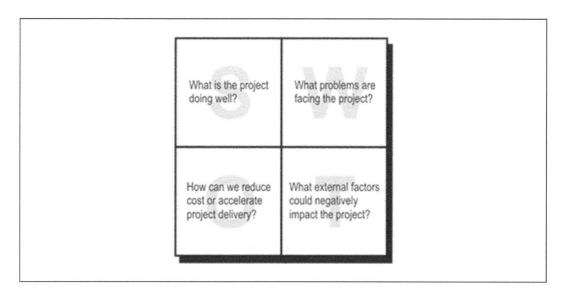

Figure 7.2   Example of a SWOT Analysis

- **Probability**—A general statement of the likelihood of the risk occurring. At this point in the process just do this by using the terms: high (greater than 75 percent), medium (approximately 50 percent), and low (less than 25 percent). This will be further elaborated upon in the qualitative and quantitative risk analysis phase.

- **Impact description**—A brief statement of the anticipated impact. At this level, the statement need not be quantitative in nature. This will be further elaborated upon in the qualitative and quantitative risk analysis phase.

- **Responsibility**—Indicate who is responsible for managing the risk.

- **Response strategy**—A place to indicate what strategy will be used to respond to the risk. This area will be filled in during the risk response planning phase.

- **Status**—A place to indicate the current status of this risk and note any changes. This area will be filled in during the risk monitoring and control phase.

This is commonly referred to as a risk register. An example of a risk register is included in Figure 7.3. There are many variations of this theme, but they generally look similar to what is shown in the figure.

During a risk identification meeting it is important to at least get a general idea as to what the potential impacts and probabilities are; however, it may be necessary to hold a separate session later in order to allocate more time to this. The idea here is to develop enough information so you can at least start to characterize the magnitude of the risks. Many practitioners like to reorder the risks from largest to smallest and divide them into threats and opportunities. This is a good approach because it will help keep the team's focus on the big issues in a place where they are likely to receive more attention.

US 41 Risk Assessment - Segment 1

Risk Category	Risk Number	Risk Type	Risk Name	Description (S.M.A.R.T.)	Risk Symptoms (or Trigger)	Affected Activity ID	Pre-Workshop Risk Data								
							Probability	Cost (in millions $)				Schedule (in months)			
								Min	Most Likely	Max	Expected Value	Min	Most Likely	Max	Expected Value
Utility/Facility Disturbances	3.1	Construction	Utility Conflicts	Utility facilities may be unsurveyed and in conflict with construction due to inaccurate or incomplete survey locates. Location of underground utilities not marked in field or markers are removed. Contractor may disturb marked utilities.	Utility facilities are unearthed during construction.		75%	$0.00	$1.00	$3.00	$0.88	0.00	0.50	3.00	0.63
Traffic during Construction	5.1	Construction	Lambeau / Oneida Traffic	There are issues associated with the August 1 deadline and the start of the NFL season. The facilities need to be able to accommodate the traffic.			30%	$0.20	$0.75	$2.00	$0.26	0.00	0.00	0.00	0.00
Permits/Agreements	9.5	Design	404 Permit Documentation Approval	Mitigation site plan documentation approval	COE denies mitigation plan		10%				$0.00	0.25	2.00	4.00	0.20
Permits/Agreements	9.9	Design	Local Agreements	This is risk related to scope pertaining to local agreements. Scope decisions need to be made and local agreement committing to the scope needs to occur. This could delay getting 1078s out.			25%					1.00	3.00	12.00	1.04
Stormwater / Water Quality	10.1	Design	Wildlife Hazard Assessment	Need to balance the need for stormwater management/need for ponds with FAA and USDA guidance to avoid/minimize potential for wildlife hazards within 5 miles of Austin Straubel Airport. This is a cost risk (Med/High)	FAA/USDA/Airport and WisDOT/DNR cannot come to agreement on pond locations/sizes or mitigation requirements.		20%	$0.10	$0.25	$0.50	$0.05				0.00
Stormwater / Water Quality	10.2a	Design	401 WQC- stormwater	Stormwater mgt. plan showing adequate TSS removal is needed prior to DNR issuing WQC. No WQC means you do not have a valid 404. Corridor analysis of stormwater management and wetland mitigation site development is behind schedule. This could potentially delay all projects, with the primary concern being the earlier LETs.	Denial of WQC		50%				$0.00	1.00	3.00	24.00	3.08
Stormwater / Water Quality	10.2b	Construction	Construction impacts of Stormwater	The construction cost impacts from a permit denial will result in cost to do mitigation			50%	$0.50	$1.00	$5.00	$0.79	1.00	3.00	12.00	

Figure 7.3 Example of a Risk Register

The risk register is probably the most important tool that a risk manager will have at his or her disposal in managing project risk. It is intended to serve as a "living" document and therefore requires constant attention. Projects are dynamic—new risks come and old ones change or go away. This document should be used to stay organized and keep pertinent information up to date. It is recommended to use version numbers and circulate revised copies of it to key project members as updates occur. Any version should record the date of the last update.

Qualitative Risk Analysis

It is not possible, nor is it desirable, to identify every conceivable risk to a project. Qualitative risk analysis is a process designed to help sort the wheat from the chaff. Time is always valuable in the world of project management, and seldom is there sufficient time and resources available for the many tasks facing project teams, let alone for performing risk analysis. Qualitative risk analysis is often the only type of risk analysis performed on projects, and usually, it is sufficient. For larger and/or more complex projects, qualitative risk analysis serves as a kind of preparatory phase for quantitative risk analysis, which is a primary focus of this book.

Qualitative risk analysis is concerned with achieving the following objectives:

- Evaluating the risks discovered during risk identification and selecting those that need careful management
- Developing a better understanding of the potential impacts of a risk
- Developing a better understanding of the probabilities of the risk's impacts occurring

Vetting the Risk Register

An initial vetting of the original risk register should be performed as the first activity in qualitative risk analysis. After the initial risk identification session there will probably be a lengthy list of potential project risks. The risk manager, with possible involvement from other team members, should initiate the process of vetting the risk register. Risks should be considered with respect to their general risk category, severity of impact, and probability. Often, risks have been identified that are similar to others—in such cases, it is recommended that these risks be combined. In other cases, there may be many small risks that could be rolled up, or *bucketed*, into a single large risk. For example, there could be multiple minor technical risks relating to a project's geotechnical conditions. Rather than treating these separately, it may be more efficient to group these into a single risk titled *Geotechnical Risks*. It is important to note that the smaller individual risks need not be lost, they are simply considered as a group for the purposes of risk management.

It is possible that there will be very unlikely but very severe risks. These risks border upon the domain of Black Swans (at least within the context of the project). They are very rare but have huge implications with respect to a project. Such risks are often systemic in nature and include events such as natural disasters and other force majeure type events as well as economic events such as market booms and busts (i.e., stock market crashes, real estate booms, and so forth) and political events (i.e., loss of funding, shifts in project scope due to policy changes, and the like). Such events are largely unmanageable and uncontrollable; however, they must still be considered though probably not quantitatively modeled.

In summary, the idea here is to develop a consolidated and concise risk register by removing redundancy and better facilitating their analyses, whether this is qualitative or quantitative.

POSTWORKSHOP ACTIVITIES

Run the RBE Model

The risk lead assembles the workshop's tangible outcomes into a statistical model that usually employs the Monte Carlo method. The statistical analysis determines the collective impact of the overall interactions between base and risks for the entire project, as a system, and produces estimates for the cost and schedule that are expressed through distributions.

The next step comprises rigorous quality control of the analysis results. The QC must be performed by an expert other than the modeler. The results are then sent to the cost lead and project manager for a common sense test. If the results pass this test, the risk lead incorporates them into the workshop report.

Workshop Report

The workshop report documents the results, basis of estimate, assumptions validated or considered as events, constraints included on analysis, project flowchart, uncertainties, risks, and workshop contributors. The report is written in support of the project team's risk management and project delivery efforts. Report preparation is a collaborative effort primarily between the project team and the cost-risk team, with final control of editing and publishing of the report resting in the hands of the risk lead. Table 7.2 provides a guide regarding roles and responsibilities for report writing.

The draft report is typically submitted within one week after the workshop. The report must be in its final form from the writer's point of view. For example, if the report's owner does not have any comments the draft report becomes final. These are, of course, guidelines and every organization will have its own timetable and report submission requirements.

When the report owner requires changes on report, the report writer must evaluate them and decide if the changes are legitimate or not. The report must reflect and represent the workshop outcomes and it is the report writer's responsibility and duty of guarding the integrity of RBE results. The workshop coordinator must review the content of the report and make sure that the results presented reflect the assumptions and constraints and other project specifics discussed in the workshop.

In the case of significant flaws with the report due to model errors, the report must be rewritten by an independent party under the condition of using the workshop findings. In case of discomfort with the workshop findings, a new process must be initiated with a new set of assumptions and constraints.

TABLE 7.2   Workshop Report Responsibilities

Project Manager	The project manager owns the report.
Subject Matter Experts	The SMEs provide written statements that express their opinions and they offer objective input during the workshop.
Risk Lead	The risk lead writes the report and makes sure that all of the pertinent information provided by others is accurate and included in the report. QA/QC is an important part of report delivery.
Cost Lead	The cost lead is responsible for preparing their portion of the report and must fully collaborate with the report writer.
Workshop Coordinator	The workshop coordinator reviews the report for completeness.

Postmitigation Activities

Usually the project manager takes the information contained in the workshop report and starts the formal process of risk management as discussed in Chapters 5 and 6. The risk management process creates a different risk mesh for the project when some risks may be terminated (retired) and some may be mitigated, while new risks may be identified and quantified.

Risk management usually requires a change in the assumptions and constraints as well as changes in the base cost and schedule. If these changes and the modification of the risk mesh are significant, a new model has to be created and run. Risk management is usually the project manager's responsibility and the results obtained considering the effects of risk planning provide data for budget and schedule assessment.

The postmitigated activities may include one or more cost risk analyses depending on project specifics and controversy. A good practice is reevaluation of the project status periodically of when significant changes occur (scope, delivery method, and schedule).

EVERYONE HAS DUTIES

Risk-based estimating is a collaborative effort when internal and external professionals work together on establishing the project cost and schedule boundaries. While collaboration is one of the key attributes that describes the process, the *systematic approach* is the second attribute of the process. The workshop coordinator provides oversight to the process and is making sure that the systematic approach is respected. He or she makes sure that all of the participants are familiar with the process and know what is expected of them.

There are four types of workshop participants besides the workshop coordinator:

1. Project manager—is the owner of the workshop report and provides resources for all activities involving RBE.
2. Risk lead—conducts the risk elicitation phase and is ultimately responsible of report writing.
3. Cost lead—validates the assumptions and base cost and schedule.
4. SMEs—provide expert advice on their area of expertise.

The next few paragraphs explain specific duties that each type of participant has to perform in order to ensure that the process goes smoothly and efficiently.

Project Manager — Duties

Project managers typically consider the RBE workshop a tool to help them improve the accuracy, consistency, and confidence in their project cost and schedule estimates. They also appreciate the project's risks information that boosts their project risk management efforts. Recognizing these benefits the project managers have vested interest on the successful outcome of the RBE. The RBE cannot happen without full commitment of the project manager.

The project manager provides resources to support the entire process. He or she makes available key SMEs who can represent the project. The project manager and workshop coordinator decide about the needs of external subject matter experts and the project manager secures their participation.

Examples of areas needed to be covered with internal and external subject matter expertise follow:

- Project management (to provide project context and relationship with stakeholders)
- Engineering (design and construction)
- Cost estimating
- Scheduling
- Environmental permits, processes, and mitigation

The project manager must ensure that the following items are available at the workshop and during the preworkshop activities:

- Participant contact information
- Project documents, aerial photos, concept plans, design drawings, illustrations, public information documents, memorandums of understanding, geo-tech info, and so forth
- Projects with multiple alternatives have to provide a description of each of the alternatives in detail that may allow planning the workshop priorities
- Current cost estimates

- A preliminary project flowchart showing key tasks and relationships from current status through completion of construction
- Current design and construction schedule, including description of how durations were determined and an explanation of the construction sequencing strategy
- Other relevant documents for the subject project

Risk Lead — Duties

The risk lead participates in the analysis of the project scope, schedule, and cost estimate to evaluate their quality and completeness while focusing on understanding the risks and uncertainties of the estimated cost and schedule. The risk lead's responsibilities are:

- Leads the risk portion of the process including risk elicitation of both threats and opportunities
- Defines the project flowchart
- Participates in cost and schedule validation and leads base uncertainty discussions
- Conducts advanced risk elicitation interviews
- Performs statistical analysis (creates and runs MCM model)
- Provides workshop reports and presentations

The risk lead plays a vital role to ensure the analysis is both sound and objective. It is also imperative that the analysis process and results are clear and usable by the project team. The process, as described in Chapter 3, must include the underlying assumptions and constraints of the analysis in a manner that is easily comprehended by the project team who will have to communicate the result of the workshop to others. The report should "tell the story" of the project scope, schedule, and cost estimate.

Cost Lead — Duties

The cost lead conducts cost and schedule validation of the RBE. Estimating is a process that is incorporated in the project development as described in Chapter 2. Therefore, there is always a story behind the estimate; it is rarely a straightforward linear process. It is imperative that the cost lead understand how the estimate was generated. The cost lead must understand the history of the estimating process. Considering the estimate history, the cost lead assists with the workshop process by taking primary responsibility for the following functions:

- Leads the review and validating of the base cost and schedule estimate effort
- Supports the project team in making any adjustments to the base estimate as a result of the review

- Participates in the development of a risk register
- Distributes the base cost against the activities identified in the flowchart
- Confirms concurrence of the validated estimate with the project team and subject matter experts
- Provides a written report on the base cost and schedule validation for inclusion on workshop report

Subject Matter Experts — Duties

External and/or internal SMEs participate in peer-level systematic project review and risk assessment to identify and describe cost and schedule risks based on the information at hand. In addition, the review process may examine how risks can be managed.

The SMEs should have extensive expertise in their specialty areas and should provide guidance and assistance on defining the cost and schedule of the project's activities related to their expertise. The SMEs should understand that risk assessment does not need to be exact to be useful and that the power of cost risk assessment workshops lies in the rigorous disciplined approach and the ability of team members to focus collectively on a broad range of topics. The SMEs should:

- Provide objective input in their field and cooperate with all team members by crossing conventional boundaries
- Have an open attitude to change by encouraging team and individuals' creative thinking
- Stay aligned to the workshop process and focus on fulfilling the workshop mission
- Have a clear understanding of the specific terminology used during workshops such as: allowances, contingency, base cost, cost uncertainty, schedule uncertainty, risk, and so forth
- Ask questions

The internal and external SMEs, working collaboratively with the workshop team, should be prepared to discuss and determine:

- Basis of estimate
- Assumptions
- Constraints
- Additional subject matter expert participation
- Authority to "de-bias" the input
- The optimal balance between effort and accuracy
- Treatment of base uncertainties

In addition to active participation in the workshops, SMEs may be asked to provide documentation of the viability of assumptions made regarding the project's configuration, scope, schedule and cost estimate, and the potential impact of risk events that may occur.

HELPFUL HINTS

- RBE is iterative in nature and represents a "snapshot in time" for a project.
- RBE normally deals with identifiable and quantifiable project-type risks—for example, those events that can occur in planning, design, bidding, construction, and changed conditions.
- It is good to remember that risk-based estimating does not provide an "answer book" with all uncertainty removed from the project. Risk-based estimating, by introducing the elements of project uncertainty and project risk, does not add costs to a project—it reveals them.
- Emphasize the importance of identifying both threats and opportunities when referring to risks.
- The workshop report provides information for decision makers to act upon.

SAMPLE QUESTIONS TO ASK

Typical cost questions to be asked by the cost lead, risk elicitor, and SMEs:
- What is the basis of the estimate?
- Does the current scope of the work match the scope that the estimate is based on?
- How current is the estimate?
- Do unit prices correlate to similar scope projects in the area? Are they truly comparable?
- Does the estimate include engineering, engineering services during construction, construction management services?
- What contingencies are built into the estimate?
- Has a change order allowance been built into the estimate?
- What is the stage of the design?
- What is the accuracy of the survey data?
- What field investigations have been done?
- What geotechnical work has been done to date? Is there data from past projects in the area?
- Cuts and fills: What has been assumed for reuse, import, export and disposal, temporary stockpiling, haul distances, location of imported materials?

- What are assumptions on stability of cuts, sheeting, retaining walls, slope protection during construction?
- If dewatering is required, are there perched water tables and other maintenance of excavations during construction? Is the treatment of dewatering required to meet permits?
- How current are surveys and estimates of use of real estate cost? Partial or full parcels acquisition?
- What is status of temporary utilities, staging areas, construction logistics . . .?
- Is there full knowledge of utilities in project area, relocation requirements, ability to isolate and shutdown? Are replacements needed prior to isolation? Can replacements be installed at proper elevation?
- What is the plan for erosion protection?
- Are there special conditions: extraordinary staffing requirements, night work, stop times due to fish or wildlife issues, noise limits, and dust control?
- What has been assumed for overhead, insurance, bonding, project management, safety, trailers, utilities, parking home office overhead, and profit?
- What are assumptions for material availability? Backfill, sheeting, piles, concrete, rebar access for delivery, double handling requirements?
- What production rates are assumed? Is this work similar to other work done in the area?
- What are assumptions for maintenance of traffic, staging of construction, needed temporary barriers, ramps, bridges, supports, technology?
- Does this project require estimated mitigation, noise walls, stormwater detention ponds, wetlands?

Typical schedule questions:

- How long have similar projects taken?
- How many $/month at average and at peak would have to be spent to meet the schedule?
- What season is it expected that the notice to proceed (NTP) will be issued? Will certain months be lost due to the start date?
- Has mobilization and demobilization time been included in the schedule? How many workers are assumed to be working on the project at the peak of construction?
- Does the construction phasing and traffic management plan match the schedule assumptions?
- How many concurrent work areas are assumed? Are there crews available to staff all of those areas?
- What are the assumed production rates for each of the major elements, earthwork, foundations, piers, beams deck, subbase, base, paving, and so forth?
- If the NTP is issued as planned, can the landscaping be completed in the required season for the specified plantings?

The preparation activities before the workshop, the workshop itself, and the analysis of the input are the main focus of RBE workshops. The project manager develops response actions for the key risks, documents the response actions, and incorporates this information into the risk management plan. The project manager tracks risks and the effectiveness of the response actions. A follow-up analysis can be performed to demonstrate the effectiveness of the response actions.

INDEX